GUIDELINES FOR MANAGING PROCESS SAFETY RISKS DURING ORGANIZATIONAL CHANGE

This book is one in a series of process safety guideline and concept books published by the Center for Chemical Process Safety (CCPS). Please go to *www.wiley.com/go/ccps* for a full list of titles in this series.

GUIDELINES FOR MANAGING PROCESS SAFETY RISKS DURING ORGANIZATIONAL CHANGE

Center for Chemical Process Safety
New York, NY

WILEY
A JOHN WILEY & SONS, INC., PUBLICATION

For general information on our other products and services please contact our Customer Care Department within the United States at (800) 762-2974, outside the United States at (317) 572-3993 or fax (317) 572-4002.

Wiley also publishes its books in a variety of electronic formats. Some content that appears in print, however, may not be available in electronic formats. For more information about Wiley products, visit our web site at www.wiley.com.

Library of Congress Cataloging-in-Publication Data:

Guidelines for managing process safety risks during organizational change / Center for Chemical Process Safety, New York, NY.
pages cm.
Includes bibliographical references and index.
ISBN 978-1-118-37909-7 (hardback)
1. Chemical industry—Safety measures. 2. Chemical industry—Management. 3. Organizational change—Health aspects. 4. Industrial safety. I. American Institute of Chemical Engineers. Center for Chemical Process Safety.
TP150.S24G857 2013
363.11—dc23 20112035829

Printed in the United States of America.

10 9 8 7 6 5 4 3 2 1

It is sincerely hoped that the information presented in this document will lead to an even more impressive safety record for the entire industry. However, the American Institute of Chemical Engineers, its consultants, the CCPS Technical Steering Committee and Subcommittee members, their employers, their employers' officers and directors, and ioMosaic Corporation and its employees do not warrant or represent, expressly or by implication, the correctness or accuracy of the content of the information presented in this document. As between (1) American Institute of Chemical Engineers, its consultants, CCPS Technical Steering Committee and Subcommittee members, their employers, their employers' officers and directors, and ioMosaic Corporation and its employees and (2) the user of this document, the user accepts any legal liability or responsibility whatsoever for the consequences of its use or misuse.

CCPS dedicates this first edition of *Guidelines for Managing Process Safety Risk During Organizational Change* to Bob G. Perry. CCPS' longest serving staff member, Bob served as the second Executive Director of CCPS from 1993–1997, and since then continues to provide support and direction as a staff consultant. Bob's leadership and operational skills honed during a long career with Union Carbide Corporation helped CCPS in its efforts to build from its early base, and resulted in the development and deployment of many new supporting tools and guidelines as well as the mentoring and development of both members and staff. Bob, along with his wife Gayle, became the historians of the organization, serving as the official CCPS archivist and photographer for many years. We thank Bob for his contributions to CCPS throughout the years and for his continued support and commitment to Process Safety, and remember Gayle fondly.

CONTENTS

List of Tables *xiii*
List of Figures *xv*
Files on the Web Accompanying This Book *xvii*
Acronyms and Abbreviations *xix*
Glossary *xxiii*
Acknowledgements *xxv*
Preface *xxvii*

1. INTRODUCTION AND SCOPE **1**

1.1 Case Study: Hickson and Welsh Ltd.—England (1994) 1
1.1.1 Lesson Learned 2
1.2 Introduction 3
1.3 The Need for Management of Organizational Change 5
1.4 Organization of this Book 6
1.5 A History of Organizational Change Management 11
1.6 Definitions Related to Management of Organizational Change 16
References 18

2. CORPORATE STANDARD FOR ORGANIZATIONAL CHANGE MANAGEMENT **21**

2.1 Case Study: BP—Grangemouth, Scotland (2000) 21
2.2 OCM Background 24
2.3 Management Commitment 25
2.4 OCM Policy 26
2.5 OCM Workflow 27
2.6 OCM Procedure 28
2.7 Definition of Organizational Change 29
2.8 Roles and Responsibilities 32
2.9 Initiate an Organizational Change 32
2.9.1 Example OCM Case 33
2.10 Review the Change 34
2.11 OCM Risk Assessment 35
2.11.1 Preparation—Selecting the OCM RA Team 36
2.11.2 Preparation—Gathering Relevant Data 37

2.11.3 Preparation—Selecting the OCM RA Method(s) and Tool(s) 39
2.11.4 Facilitation of the Risk Assessment 52
2.11.5 Documenting the Risk Assessment 54
2.12 Action and Implementation/Transition Plans 55
2.12.1 Example OCM Case 58
2.13 Postimplementation Monitoring 59
2.13.1 Example OCM Case 60
2.14 Closeout 61
2.15 Conclusion 61
References 62

3. MODIFICATION OF WORKING CONDITIONS 65

3.1 Case Study: Esso—Longford, Victoria, Australia (1998) 65
3.1.1 Lessons Learned 67
3.2 Modifying Location, Communications, or Time Allocation for People 68
3.3 Case Study: Changes in Shift Schedules and Staffing During Turnarounds 69
3.3.1 Lessons Learned 71
3.4 Changes to Terms and Conditions of Employment (e.g., Hours, Shifts, Allowable Overtime) 72
3.5 Staffing During Turnarounds, Facility-Wide Emergencies, or Extreme Weather Events 74
3.6 Impacts and Associated Risks 76
3.7 Special Training Requirements 79
3.8 Conclusion 80
References 81

4. PERSONNEL CHANGES 83

4.1 Case Study: Union Carbide—Bhopal, India (1984) 83
4.1.1 Lessons Learned 85
4.2 Case Study: Bayer CropScience, LLC—Institute, West Virginia, USA (2008) 87
4.2.1 Lessons Learned 88
4.3 Changes in Plant Management (e.g., Plant Manager or EHS Manager) 91
4.4 Replacement of a Subject Matter Expert 92
4.5 Replacing the Incumbent in a Position that Directly Affects Process Safety 93
4.6 Strikes, Work Stoppages, Slowdowns, and Other Workforce Actions 93
4.7 Emergency Response Team Staffing 95
4.8 Impacts/Associated Risks 95

4.9 Organizational Change Procedures versus OCM for New Hires, Promotions, etc. 97
4.10 Conclusion 98
References 98

5. TASK ALLOCATION CHANGES 99

5.1 Downsizing Examples 99
5.2 Task Allocation Changes 101
5.3 Job Competency Change 102
5.4 Case Study: Bayer CropScience LLC—Institute, West Virginia, USA (2008) 103
5.4.1 Lessons Learned 104
5.5 Assigning New Responsibilities 105
5.6 Temporary Backfilling 106
5.7 Vanishing Task Allocations 106
5.8 Case Study: BP—Whiting, Indiana, USA (1998–2006) 107
5.8.1 Lessons Learned 108
5.9 Impacts/Associated Risks 109
5.10 Conclusions 111
References 112

6. ORGANIZATIONAL HIERARCHY CHANGES 113

6.1 Centralization or Decentralization of Job Functions 114
6.2 Case Study: Esso—Longford, Victoria, Australia (1998) 115
6.2.1 Lessons Learned 115
6.3 Reorganizations and Delayering the Hierarchy 117
6.4 Impacts/Associated Risks 119
6.5 Changes to Span of Control 121
6.6 Impacts/Associated Risks 122
6.7 Linear vs. Matrix Organization 122
6.8 Case Study: BP—Texas City, Texas, USA (2005) 124
6.8.1 Lessons Learned 125
6.9 Impacts/Associated Risks 126
6.10 Acquisitions, Mergers, Divestitures, and Joint Ventures 127
6.11 Case Study: Anonymous, USA (1998) 127
6.11.1 Lessons Learned 128
6.12 Associated Risks 128
6.13 Case Study: Union Carbide—Bhopal, India (1984) 129
6.13.1 Lessons Learned 129
6.14 Changing Service Providers 132
6.15 Impacts/Associated Risks 132
6.16 Conclusion 133
References 134

7. ORGANIZATIONAL POLICY CHANGES 135

7.1 Case Study: Dupont—Delaware, USA (1818) 135
7.1.1 Lessons Learned 136
7.2 Changes to Mission and Vision Statements 136
7.3 New and Revised Corporate Process Safety Related Policies/Procedures 138
7.4 Major Changes to Policy or Budgets for Maintenance or Operations 139
7.5 Impacts/Associated Risks 140
7.6 In/Outsourcing of Key Departmental Functions (e.g.., Engineering Design or Maintenance) 142
7.7 Staffing-Level Policy Changes (Shutdowns, Turnarounds, Startups) 144
7.8 Special Training Requirements 146
7.9 Conclusion 146
References 147

APPENDIX A: EXAMPLE TOOLS FOR EVALUATING ORGANIZATIONAL CHANGES 149

A.1 What-If Analysis 149
A.2 Checklists 150
A.3 Other Risk Assessment Tools 188
A.4 Special Competency Assessment for Control Room Staff 188
References 198

APPENDIX B: EXAMPLE PROCEDURES FOR MANAGING ORGANIZATIONAL CHANGES 199

INDEX 237

LIST OF TABLES

Table 1.1.	A Selection of OCM-Related Incidents	9
Table 1.2.	7S Model – Organizational Aspects	17
Table 2.1.	Comparison of MOC and OCM Workflow Tasks	27
Table 2.2.	OCM RA Methods and Tools Comparison	40
Table 2.3.	Activity Mapping Form 1 “Old Role”	48
Table 2.4.	Activity Mapping Form 2 “New Role”	49
Table 3.1.	A Potential Atypical Scenario and Mitigation Options	77
Table 4.1.	Potential Risks Associated with Personnel Turnover	96
Table 5.1.	Examples of Disturbances, Accidents, and Major Costs Where Downsizing Has Played a Major Role (Sweden)	100
Table 5.2.	Task Allocation Change Impact What-Ifs	109
Table 5.3.	Examples of HAZOP Parameters for Organizational Change Review	111
Table A.1.	Management Leadership, Commitment, and Accountability	153
Table A.2.	Risk Assessment and Management	157
Table A.3.	Facilities Design and Construction	158
Table A.4.	Operations and Maintenance	160
Table A.5.	Management of Change	163
Table A.6.	Information/Documentation	164
Table A.7.	Personnel and Training	165
Table A.8.	Third-Party Services	168
Table A.9.	Incident Investigation and Analysis	169
Table A.10.	Community Awareness and Emergency Preparedness	171
Table A.11.	Operations Integrity Assessment and Improvement	174
Table A.12.	Process Safety	175
Table A.13.	Generic What-If	176
Table A.14.	Example Checklist for Organizational Management of Change	177
Table A.15.	Example MOC Organizational Change Checklist	179
Table A.16.	Example Task List	185
Table A.17.	Ladder for Management Procedures	196
Table B.1.	Operations and Safety Effectiveness Checklist	207
Table B.2.	Safety and Health Management Checklist	208
Table B.3.	Safe Work Practices Checklist	211

Table B.4.	Process Safety Management/Risk Management Program (PSM/RMP) Checklist	213
Table B.5.	Contractor Safety Checklist	214
Table B.6.	Emergency Response Checklist	215
Table B.7.	EH&S Regulatory Compliance Checklist for Selected Regulations	216
Table B.8.	Occupational Health Checklist	218
Table B.9.	Process Unit Operability and Safety Effectiveness Checklist	220
Table B.10.	Management of Personnel Change (MOPC) Checklist	224

LIST OF FIGURES

Figure 2.1.	Example Work Flow for OCM Procedure	31
Figure 3.1.	A Simplified Schematic of the Demethanizer Column and Associated Equipment	71
Figure A.1.	Flow Chart of the Assessment Method for Documenting Work Arrangement	193

FILES ON THE WEB ACCOMPANYING THIS BOOK

Access Managing Process Safety Risks During Organizational Change tools and documents using the Microsoft Explorer web browser at:

http://www.aiche.org/ccps/publications/OrgChangeMaterial.aspx

Password: ORGCHANGE2012

ACRONYMS AND ABBREVIATIONS

ACC	American Chemistry Council
AIChE	American Institute of Chemical Engineers
AIM	Automated Information Management
ANSI	American National Standards Institute
API	American Petroleum Institute
API RP	American Petroleum Institute Recommended Practice
BP	British Petroleum
CCPS	Center for Chemical Process Safety
CERT	Community Emergency Response Team
CFR	Code of Federal Regulations
CHIS	Chemical Information Sheet
CPI	Chemical Process Industries
CSB	U.S. Chemical Safety and Hazard Investigation Board
DAFWC	Days Away From Work Case
DCS	Distributed Control System
EBC	Equipment Basic Care
EHS	Environmental, Health, and Safety
EMT	Emergency Medical Team
ERP	Emergency Response Plan
ERT	Emergency Response Team
FCCU	Fluid Catalytic Cracking Unit
HAZCOM	Hazard Communication
HAZID	Hazard Identification
HAZOP	Hazard and Operability
HPI	Hydrocarbon Process Industry
HSE	Health, Safety, and Environmental (refer to EHS)
HSSE	Health, Safety, Security, and Environmental
IChemE	Institution of Chemical Engineers
I/E	Instrument and Electrical
IMT	Incident Management Team
JIT	Just in Time
KPI	Key Performance Indicator
LDAR	Leak Detection and Repair
LOPA	Layer of Protection Analysis
MAWP	Maximum Allowable Working Pressure

MERT	Medical Emergency Response Team
MIBK	Methyl Isobutyl Ketone
MIC	Methyl Isocyanate
MMS	Maintenance Management System
MOC	Management of Change
MOOC	Management of Organizational Change
MOPC	Management of Personnel Change
MOV	Mechanically Operated Valve
OCM	Organizational Change Management
OMS	Operating Management System
OSHA	Occupational Safety and Health Administration
PHA	Process Hazard Analysis
PLC	Programmable Logic Controller
PM	Preventative Maintenance
PPE	Personal Protective Equipment
PSM	Process Safety Management
PSSR	Pre-Startup Safety Review
PSI	Process Safety Information
PSV	Pressure Safety Valve
QC	Quality Control
R&D	Research and Development
RA	Risk Assessment
RBI	Risk-Based Inspection
RCRA	Resource Conservation and Recovery Act
RMP	Risk Management Plan
RRE	Roles, Responsibilities, and Expectations
SARA	Superfund Amendments and Reauthorization Act
SHE	Safety, Health, and Environment (refer to EHS)
SIL	Safety Integrity Level
SIS	Safety Instrumented System
SME	Subject Matter Expert
SOP	Standard Operating Procedure
SPCC	Spill Prevention, Control, and Countermeasures
SRT	Safety Response Team
TSC	Technical Steering Committee
UCIL	Union Carbide India Limited

GLOSSARY

Activity mapping
A process by which activities required to accomplish a particular objective are evaluated and assigned to individuals or roles within the organization.

Bow tie
A method used to graphically represent a hazard and the possible causes and outcomes related to that hazard.

Management of change (MOC)
A system to identify, review, and approve all modifications to equipment, procedures, raw materials, and processing conditions—other than "replacement in kind," prior to implementation.

Organizational change
Any change in position or responsibility within an organization or any change to an organizational policy or procedure that affects process safety.

Organizational change management
A method of examining proposed changes in the structure or organization of a company (or unit thereof) to determine whether the changes introduce new hazards or increase the risk to employee health and safety, the environment, or the surrounding community.

Process safety management (PSM)
A program or activity involving the application of management principles and analytical techniques to ensure the safety of process facilities. Sometimes this is also called process hazard management, safety engineering, or technical safety. Each principle is often termed an "element" or "component" of process safety.

Risk assessment
The assessment of risk presented by a change. Considers all possible outcomes and their significance, as well as the likelihood of realizing those outcomes.

Task mapping
A process by which work tasks are evaluated and assigned to certain individuals or roles within the organization.

ACKNOWLEDGEMENTS

The American Institute of Chemical Engineers (AIChE) and the Center for Chemical Process Safety (CCPS) express their appreciation and gratitude to all members of the Organizational Change Management Project and their CCPS member companies for their generous support and technical contributions in the preparation of these *Guidelines.* The AIChE and CCPS also express their gratitude to the team of writers from ioMosaic Corporation.

SUBCOMMITTEE MEMBERS:

John Wincek	Committee Chairman, Croda
Habib Amin	Contra Costa County Health Service
Steve Arendt	ABS Consulting
Mike Broadribb	Baker Risk
Glen Crowe	Potash Corp
Jim Conner	Celanese (Retired)
Ken Harrington	Chevron Phillips
Don Lanier	Bayer MaterialScience, LLC
Jim Miller	ConocoPhillips
Keith Pace	Praxair
Cathy Pincus	ExxonMobil
Fran Schultz	SABIC
David Thaman	PPG
Rachel Vincze	Suncor
Dave Belonger	Staff Consultant, CCPS

CCPS wishes to acknowledge the many contributions of the ioMosaic Corporation staff members who prepared this book, especially the principal writer Molly Myers.

Before publication, all CCPS books are subjected to a thorough peer review process. CCPS gratefully acknowledges the thoughtful comments and suggestions of the peer reviewers for this book, enhanced its accuracy and clarity.

PEER REVIEWERS:

Troy Bennett	Arizona Chemical
Louisa A. Nara	CCPS
Jonas Duarte	Chemtura

David Cones	ConocoPhillips
David Guss	Nexen, Inc.
Enrique Elias Troy	Praxair, Inc.
Mark Kleis	Praxair, Inc.
Kim Mullins	Praxair, Inc.
Dale Dressel	Solutia, Inc.

CCPS does not ask peer reviewers to endorse the book and does not show them the final draft before release.

PREFACE

The American Institute of Chemical Engineers (AIChE) has been closely involved with process safety and loss control issues in the chemical and allied industries for more than four decades. Through its strong ties with process designers, constructors, operators, safety professionals, and members of academia, AIChE has enhanced communications and fostered continuous improvement of the industry's high safety standards. AIChE publications and symposia have become information resources for those devoted to process safety and environmental protection.

AIChE created the Center for Chemical Process Safety (CCPS) in 1985 after the chemical disasters in Mexico City, Mexico, and Bhopal, India. The CCPS is chartered to develop and disseminate technical information for use in the prevention of major chemical accidents. The center is supported by more than 150 chemical process industries (CPI) sponsors who provide the necessary funding and professional guidance to its technical committees. The major product of CCPS activities has been a series of guidelines to assist those implementing various elements of a process safety and risk management system. This book is part of that series.

Management of change (MOC), including organizational change management (OCM), is a fundamental element of successful process safety programs. However, facilities continue to be challenged to maintain successful MOC and OCM programs in a way that improves total process safety over time. The CCPS Technical Steering Committee initiated the creation of these guidelines to assist facilities in meeting this challenge. This book contains approaches for designing, developing, implementing, and continually improving an organizational change management program.

1

INTRODUCTION AND SCOPE

It has long been acknowledged that when not properly evaluated and controlled, changes in physical equipment in a facility can lead to serious incidents with potentially severe consequences. Management-of-change (MOC) systems, replete with a variety of electronic systems, flow charts, and checklists, have been developed by a number of reliable organizations throughout the world to deal with these physical changes. However, other types of changes such as changes in job responsibilities, loss of key personnel, or even changes in shift hours may not be included in an MOC program. It is less well understood that these and other nonphysical changes, collectively referred to as "organizational changes," can also lead to serious incidents with potentially severe consequences. Due to their focus on managing physical changes, most MOC systems have overlooked or only superficially addressed organizational change management and the impact of organizational changes that affect process safety. Although there are many types of organizational changes that a company can make, the focus of this book is on changes that may affect process safety. When the generic term organizational change management (OCM) is used throughout this text, keep in mind that it only refers to those changes which may affect process safety.

1.1 CASE STUDY: HICKSON AND WELSH LTD.—ENGLAND (1994)

On the afternoon of September 21, 1992, a jet of flame erupted from a manway on the side of a batch still at the factory of Hickson & Welch Ltd., Wheldon Road, Castleford, West Yorkshire, England.

A total of 5 people were killed and another 17 were injured, in addition to over 100 reports of toxic effects.

This incident happened during the cleanout of the "60 Still Base," which contained a sludge rich in dinitrotoluenes and nitrocresols. (These compounds can be explosive in the presence of strong alkali or strong acid and have also been known to explode when exposed to heating alone under certain conditions.) This vessel had not been cleaned out during the 30 years that it had been in service. Prior to the cleaning, the sludge had been heated using steam coils built into the still, and instructions were given to not let the temperature of the sludge exceed 90° C. Unfortunately, the only temperature probe in the still was not in contact with the sludge.

In August 1991, the management structure of the Fine Chemicals Division of Hickson International, plc was reorganized. The structure changed from a linear structure to a matrix in which the role of plant manager was eliminated. Instead, the plant was managed through coordination of senior operatives who were appointed to act as team leaders.

1.1.1 Lesson Learned

The reorganization resulted in the area manager being overloaded and unable to provide the attention necessary to properly plan the cleanout of the 60 Still Base. Although there was evidence that some technical people within the company were aware of the potential for self-heating of the nitrotoluenes, this information was not available to or considered by the people planning this cleaning operation.

The organizational change at Hickson & Welch left them vulnerable to this process safety incident. There was some evidence of a loss of corporate knowledge when people changed positions. The area manager was now responsible for maintenance activities, which was a new role responsibility, and the workload was not properly balanced to allow adequate time and attention to process safety issues.

This is just one of the types of organizational changes that will be covered in this book. As you will see, various types of organizational changes have the potential to be contributing factors in process safety incidents. It is important to understand these impacts and associated risks prior to implementing any organizational change and develop an action plan to reduce these risks.

1.2 INTRODUCTION

OCM in particular has often been overlooked by many guideline publications in the past. Documents have historically disregarded the topic, mentioned it in passing, or focused on only a few of its aspects. As a result, major decisions regarding reduction in staffing levels, reorganization of the corporate hierarchy, modifications to shift schedules, or adjustment of personnel responsibilities can often be finalized by individuals or committees who lack a full understanding of how these changes may affect process safety and, by extension, the health and safety of their employees, contractors, or the surrounding community.

It has been well understood that physical changes can have an adverse impact on process safety, hence the utilization of a management-of-change process. However, it may not be as clear how changes to an organization can impact process safety. The purpose of this book is to provide an understanding of how organizational changes could potentially lead to process safety incidents, even though the contribution of the organizational change may not be as obvious as a physical change. The book will include case studies of actual incidents along with more generic examples and discussions of a range of changes that should be evaluated.

Safety-critical positions may be affected depending on the type of change involved. Responsibilities and roles may change at a number of different levels of the organization, resulting in a breakdown in the typical system of checks and balances. Essential duties can be neglected without a comprehensive approach to evaluating, approving, and documenting these changes. The lack of an OCM system, or the existence of a flawed one, has been found to be a contributing factor and/or a root cause in a number of disastrous incidents at industrial facilities in recent decades.

To establish an effective OCM process, it is essential to start with top decision-makers, whose visible enthusiasm is required if a true commitment to safety is to be accepted and upheld by their employees. When the highest echelons of the corporate structure take an active role in seeing that OCM goals are accomplished, a successful process safety management (PSM) program can be improved by adding this important component.

OCM should include veteran personnel by recording their anecdotal knowledge of their responsibilities before they retire or move on so that future employees have access to this invaluable, and often undocumented, insight. An employee of 35 years remembers the locations of floor drains, long-ago abandoned and forgotten, that could cause environmental releases if loss of containment occurs in his unit. He also knows that obscure but essential parts for the equipment installed during his first months on the job are impossible to obtain unless you contact a certain distributor and allow three months of lead time. This was why he tried to keep spares of those types of parts on hand.

The OCM process should also embrace new hires by initiating them into a corporate culture of "Safety First" on their first day of employment and by reinforcing this regularly. It should incorporate the entire workforce, whose input needs to be both respected and actively sought when reorganization of any kind occurs so that no detail is overlooked with regard to health and safety impacts.

OCM has successfully become an integrated part of the company philosophy and its overall process safety strategy when everyone within a company can navigate organizational changes smoothly without negatively affecting the safety of employees, the community, or the environment. This book should be of assistance to you on your way to achieving this goal.

1.3 THE NEED FOR MANAGEMENT OF ORGANIZATIONAL CHANGE

Organizational change is an unavoidable aspect of doing business. When an experienced employee retires, advances, or moves on to another opportunity, capturing the knowledge that employee gained through years of experience in a particular area is crucial. Downsizing often creates the need to combine functional positions. Critical responsibilities of every position should be accounted for, to maintain PSM systems when job descriptions are merged or it will not be possible to maintain full functionality of all elements of safety programs. During an unexpected snowstorm over the holidays, a temporary shifting of tasks due to vacationing personnel means there will be a need for review and documentation of procedures for employees who may be filling unfamiliar roles (as well as additional training). When a hiring freeze means a vacant position cannot be filled for weeks or months, someone should be made accountable for the critical duties of that position in the interim. As corporations acquire smaller companies they assimilate new employees and different business structures and safety cultures. Positions that appear redundant should be thoroughly evaluated throughout this process to ensure critical responsibilities are not overlooked. Any of these common events, among a myriad of other organizational changes, could result in catastrophic consequences if the changes are not successfully administered. Effective OCM procedures should include a system for managing potential modifications to all of these areas.

As with any key change, a vital step in the transition from conversations about OCM to the implementation of a practical and successful system is the initiation of senior management into the assembly of active supporters. This is typically handled by assigning someone in the management chain as a champion for this initiative who is responsible for getting buy-in throughout the organization. Health and safety specialists must sometimes walk a fine line between what is best for the safety of personnel and what is realistically accepted by upper levels of management. Introducing a new OCM program or updating and improving an existing one requires financial and personnel resources. Without an understanding of how OCM can affect their unit, facility, or overall enterprise, high-level decision-makers can end the improvement process before it begins. Existing corporate culture may not place an emphasis on OCM since its impact on health and safety may not be as obvious as other modifications. Managers must juggle resource fluctuations, health and safety issues, profit margins, product output, and a myriad of other high-priority items on a daily basis. Enabling them to understand the positive impact a flourishing OCM system can play in all these areas is key to highlighting its importance. Once senior management is committed, OCM can begin to work its way through the corporate culture and become a regular component of the overall MOC process.

1.4 ORGANIZATION OF THIS BOOK

Chapter 2 of this book will present the framework for establishing an OCM program. Chapters 3 – 7 will discuss implementation of OCM procedures for different types of changes including:

- Chapter 3: Modification of Working Conditions
  - Changes to the terms and conditions of employment such as hours, shifts, allowable overtime (including temporary deviations)
  - Modifying the location, communications, or time allocation for people
  - Staffing during extreme weather events, facility-wide emergencies, or turnarounds
- Chapter 4: Personnel Changes
  - Change of plant management [e.g., plant manager, environmental, health, and safety (EHS) manager]
  - Project changes (e.g. new project manager, replacement of a project engineer)
  - Replacing a subject matter expert (SME)/core competency [e.g., relief sizing, risk assessment, mechanical integrity, incident investigation, process hazard analysis (PHA), or hazard identification facilitation.]
  - Replacing the incumbent in a position that affects process safety (including corporate staff)
  - Strikes, work stoppages, downsizing/layoffs, retirements, slowdowns, etc.
  - Emergency response team staffing for facility-wide emergencies
- Chapter 5: Task Allocation Changes
  - Job competency requirement changes
  - Relinquishing an individual's responsibilities for tasks without those tasks being reallocated
  - Requiring individuals to take on new responsibilities demanding skills and competencies unconnected with those previously required
  - Temporarily not filling a position (e.g., hiring freeze)

  - Temporary backfill for illness, leave of absence, vacation, or temporary assignments
- Chapter 6: Organizational Hierarchy Changes
  - Centralization or decentralization of job functions (engineering, maintenance, EHS, etc.)
  - Reorganization and delayering the hierarchy
  - Changes to span of control
  - Linear versus matrix organization
  - Acquisitions, mergers, divestitures, and joint ventures
  - Changing service providers
- Chapter 7: Organizational Policy Change
  - New and revised corporate process safety related polices/procedures
  - In/outsourcing of key departmental functions such as engineering design or maintenance
  - Major changes to policy or budgets for maintenance or operations
  - Staffing level policy changes (startup, shutdown, turnaround)
  - Changes to mission or vision statements

OCM program designers may wish to include business impacts of the change such as the capacity to deliver quality products in addition to the safety impacts. However, this book will focus strictly on the process safety impacts.

Table 1.1 includes a sampling of incidents at process facilities in which a failure to correctly implement OCM was determined to be a contributing cause. These incidents will be used to illustrate particular aspects of OCM throughout this book. The chapter or chapters in which these case studies appear are listed in the far right column.

Table 1.1 A Selection of OCM-Related Incidents

Year	Company	Location	Incident	Damage	OCM-Related Causes	OCM Chapter
1984	Union Carbide	Bhopal, India	Issues arising during a pipe washing procedure resulted in pressure rise and release of toxic gas from vent stack	Est. 2,000 – 10,000 fatalities; unknown injuries	Institutional knowledge not captured prior to downsizing; employees performing unfamiliar tasks; joint venture doesn't alleviate responsibility for process safety	4, 6
1998	Anonymous*	United States	Electrical fault resulted in emergency shutdown of production unit	No injuries; emergency shutdown resulted in lost production	Change in ownership didn't adequately transition the mechanical integrity program to the new organization	6
1998	Exxon-Esso Australia	Longford, Victoria, Australia	Warm oil pump restarted after reboiler had cooled below allowable temperatures; restarting the oil pump resulted in brittle fracture of the reboiler, explosion and fire	Two fatalities; eight injuries; destruction of reboiler, two-week natural gas service disruption	Centralization and relocation of engineering staff off site from facility with no follow-up regarding maintaining communications	3, 6
1998 – 2006	BP Amoco**	Whiting, Indiana, United States	Modifications to relief systems were delayed; maintenance of relief devices was inadvertently overlooked	No damage; risks posed from overpressured equipment	Job tasks were not reassigned when a project engineer left the company, so the project "fell through the cracks"	5

*Information from CCPS Process Safety Incident Database – the company name is confidential.
** BP Amoco is now known as BP. plc.

TABLE 1.1 A Selection of OCM-Related Incidents *(Continued)*

Year	Company	Location	Incident	Damage	OCM-Related Causes	OCM Chapter
2000	BP Amoco**	Grangemouth, Scotland	Three separate incidents in May and June: electrical fault resulting in loss of power and plant shutdown; steam line rupture; and fire in the fluidized catalytic cracking unit	Plant shutdown; public road temporarily closed; fire damage to FCCU	Organizational and personnel changes made at all levels without mapping responsibilities or ongoing competency development; utilities division in charge of multiple projects became overloade d	2
2005	British Petroleum	Texas City, Texas, United States	Isomerization startup resulting in a hydrocarbon release that ignited	Fifteen fatalities; multiple injuries; significant property damage	Ineffective organizational structure	6
2008	Bayer CropScience, LLC	Institute, West Virginia, United States	A runaway chemical reaction during startup of the methomyl unit resulted in explosion, fire, and release of toxic gas	Two fatalities; eight injuries; destruction of equipment	Reduction in technical advisors without ensuring their knowledge was retained elsewhere on staff: capital project personnel unfamiliar with control system and process	4, 5
Unknown	Not specified	Not specified	Liquid overflowed from a column into the flare stack; no ignition	No injuries; failure of the flare stack	Modification of the working hours	3

*Information from CCPS Process Safety Incident Database – the company name is confidential.

** BP Amoco is now known as BP, plc.

1.5 A HISTORY OF ORGANIZATIONAL CHANGE MANAGEMENT

Practical discussions regarding how to manage organizational change have not typically occurred alongside discussions about operational or process change management. Companies have been dealing with organizational change for a long time; however, the focus for most organizational changes has been regarding the business aspects such as efficiency and profitability. Accident prevention strategies have historically focused on improving technology or reducing human error in most companies, and failures or shortcomings in management systems were largely overlooked. After several fatal and destructive accidents in process facilities in the 1970s and 1980s, governments and companies identified the lack of serviceable management systems as an underlying source of these incidents.

After the Flixborough incident in 1974, companies began to realize the importance of managing changes to equipment and processes. Loughborough University conducted a training course concerning change management in 1976. Technical journals began publishing articles regarding management of change in the mid-1970s. BP and other U.K. companies developed corporate standards for safely managing changes at their facilities.

The Center for Chemical Process Safety (CCPS) was created in 1985 as a result of these tragedies, but in particular the incident in Bhopal, India. CCPS attempted to consolidate the best practices for management of change by publishing the *Guidelines for Technical Management of Chemical Process Safety* in 1989. This book covered a wide range of changes within an organization, delineated into 12 elements of process safety management. Including only a brief mention of the need to manage organizational change, it was published a year before a new Occupational Safety and Health Administration (OSHA) PSM standard [29 Code of Federal Regulations (CFR) 1910.119] was proposed. The section concerning organizational change discusses the need to document a veteran employee's institutional knowledge before he or she departs and other issues that may be raised when significant changes to staffing levels are made. The extensive range of other organizational modifications and their inherent challenges were not identified; and no practical guidance was provided for how to approach these issues in a responsible manner.

The OSHA PSM standard "Process Safety Management of Highly Hazardous Chemicals" (29 CFR 1910.119) was promulgated in 1992. While it required MOC to be a key element of any complete PSM program, it did not include changes to the organization as a trigger for the MOC process. At this stage, the CCPS publication was still the only guideline that identified changes to organizational structure as a possible concern with regard to process safety. Within the United States, this remained the case until several years later. Additional publications within the United States and abroad have since approached various segments of organizational change management. These efforts still lacked a comprehensive overview of, and strategy to manage, the many potential organizational factors that may affect process safety on unit, plant, and corporate levels.

The American Chemistry Council (ACC), originally the Manufacturing Chemists' Association and later the Chemical Manufacturers' Association, published "Management of Safety and Health During Organizational Change" in 1998 as a resource for managers and health and safety professionals when dealing with changes in organization of their facility or an individual unit within the facility. It contains a number of detailed worksheets and checklists to help identify the potential health and safety impacts associated with proposed modifications. The effect of impending changes on either management's commitment to safety or its perceived commitment to safety is evaluated. Questions also focus on whether these suggested changes may require updates to facility-wide safety training programs or only require departmental updates. The worksheets guide management in determining whether the changes will necessitate revision of a variety of operating and permitting procedures. Potential compliance issues with PSM, emergency response, general industry, and other OSHA regulatory standards are addressed as well.

The health services department in Contra Costa County, California, released its "Industrial Safety Ordinance Guideline" in 1999, with a chapter dedicated to the management of change for organizational changes. This chapter requires companies doing business within the county to have written MOC procedures in place for permanent staffing changes, reorganization of operations, or emergency response. It outlines the requirements that must be included in an MOC procedure so that the impact of the change on health and safety can be assessed. The document recognizes that additional MOC procedures may be necessary to adequately account for other types of staffing changes. It defines the need for the formation of a "Change Team" or "MOC Team" to facilitate comprehensive management of any organizational changes in these areas. The change team is expected to include employees from engineering, operations, maintenance, and safety and health. This team is charged with fully documenting the MOC process, particularly in the event of a discrepancy between management decisions and team conclusions.

In 2001, the United Kingdom Health and Safety Executive published its "Contract Research Report 348/2001," which discusses the importance of staffing arrangements with regard to process safety. This document focused primarily on the issue of staffing requirements during emergency incidents. It includes an in-depth discussion about the effects of staffing levels on the detection, diagnosis, and correction of hazardous situations at the plant level. The study presents a two-stage strategy for analysis of these changes, incorporating both a physical assessment and a ladder assessment of proposed changes. The physical assessment entails a review of anticipated staffing arrangements against six principles, including considerations such as minimization of typical distractions in the control room and allocating emergency responsibilities to those employees who are not primarily responsible for bringing the plant back to safe operating conditions. The ladder assessment is used to examine organizational factors through the use of 12 anchored descriptive rating scales. The ladders are prepared using responses to questions that target potential concerns in categories such as alertness and fatigue, training and development, continuous improvement of safety, and willingness to initiate major hazard recovery.

In 2003, CCPS released "Retaining Corporate Process Safety Memory," a record of findings based on research that began during a Technical Steering Committee (TSC) meeting in 2001. Members of the TSC participated in a roundtable discussion during this meeting. They documented factors that could result in a loss of process safety knowledge, techniques their companies use to capture this information and pass it on to future generations, and ways CCPS could assist its members in facilitating process safety memory retention. The notes from this roundtable were used to develop a related questionnaire that was completed by attendees at additional workshops and conferences over an 18-month period. This research provided valuable insight into the causes and potential remedies for process safety memory loss at the plant, facility, and corporate levels. Survey responses routinely indicated a lack of adequate tracking of these changes as well as the need for all levels of the corporate culture to appreciate the importance of OCM. Among the suggestions received from participating members was that CCPS develop and publish a comprehensive process safety knowledge management guideline.

The Canadian Society for Chemical Engineering released its own guideline in 2004, which recognized the management of organizational change as a vital component of both a successful health and safety program and a successful business strategy. It considered multiple workforce changes that may affect process safety, provided general requirements for the management of these changes, and offered a flow chart and a risk assessment checklist to track potential concerns. It was acknowledged within this document that the process described is only one method for dealing with organizational changes. It was written to provide guidance for large, small, temporary, or permanent changes. While staffing reductions are a common occurrence, there comes a point where any additional reductions can affect unit safety. During normal daily operations this may not be a problem, but in the event of a minor incident, these reduced levels can cause difficulties to escalate quickly.

CCPS released *Guidelines for Risk Based Process Safety* in 2007. This book expanded the original 12 elements of process safety into 20 elements and included a slightly more in-depth discussion of organizational change. The additional elements were developed from experience gained over the 15 years of formal process safety implementation since their original publication in 1989.

In 2008, CCPS published the *Guidelines for Management of Change for Process Safety*. This MOC guide includes an extensive list of modifications that could increase safety risks, including examples of organizational and staffing changes. This particular guideline is geared toward the design, implementation, and continuous development of an overall MOC program and therefore does not provide an explicit study of organizational change or its effects on process safety.

CCPS issued the *Guidelines for Process Safety Acquisition Evaluation and Post Merger Integration* in 2010. This book focuses on the organizational changes inherent to mergers and acquisitions; this guideline provides several checklists to assist with the management of process safety both pre- and postmerger. It addresses due diligence items related to safety that might typically be overlooked during these proceedings.

An update was issued to the Contra Costa County "Industrial Safety Ordinance Guidelines" in 2011. This update requires the conduct of a management of organizational change for changes in permanent staffing levels and reorganizations in maintenance, health, and safety, as well as what was previously required in operations and/or emergency response. This applies to stationary sources using contractors within permanent positions in operations and maintenance. Prior to conducting the management of organizational change, stationary sources are to ensure that descriptions of job functions are complete, current, and accurate for the positions under consideration. Staffing changes that last longer than 90 days are considered permanent. Temporary changes associated with strike preparations are included in the management of organizational change requirements.

Despite all of these publications, there is still not a comprehensive resource to explain the need for an OCM program and, if so, how to implement such a program. This publication intends to fill that gap and further the continuous improvement cycle which is the hallmark of a good PSM program.

1.6 DEFINITIONS RELATED TO MANAGEMENT OF ORGANIZATIONAL CHANGE

Key terms used in this book are defined within the Glossary. Some of this terminology warrants further supplementary explanation.

Process safety management: A program or activity involving the application of management principles and analytical techniques to ensure the safety of process facilities. Sometimes this is also called process hazard management, safety engineering, or technical safety. Each principle is often termed an "element" or "component" of process safety.

While the term process safety management (PSM) is used in the United States by OSHA, the elements of process safety and the concept of managing those elements (hence process safety management) are universal and not limited to one legal jurisdiction. In this text, the term "process safety management" and the acronym "PSM" are understood to be generic and not specific references to U.S. OSHA regulations unless specifically indicated as such.

TABLE 1.2 7S Model – Organizational Aspects

Hard Aspects	Soft Aspects
Structure	Strategy
Staff	Style
Skills	Shared values
Systems	

Organizational change: Any change in position or responsibility in an organization or any change to an organizational policy or procedure that affects process safety.

These types of changes can include modification of working conditions, personnel changes, task allocation changes, organizational hierarchy changes, and organizational policy changes. Specific examples of these modifications will be explored throughout this guideline. One framework that provides a list of these types of changes is the 7S Model. See Table 1.2.

Management of change: A system to identify, review, and approve all modifications to equipment, procedures, raw materials, and processing conditions, other than "replacement in kind," prior to implementation.

It is a method for analyzing, documenting, and directing modifications or improvements to processes, equipment, organization, operating procedures, or other characteristics of a business enterprise. One purpose of MOC is to ensure that none of these changes result in the introduction of new hazardous conditions for personnel, the general population, or the environment. A thorough evaluation should always be performed so that any potential harmful effects can be mitigated or eliminated prior to implementation of a change. MOC is an essential element for any PSM system to truly be effective. Both business performance and health and safety performance can benefit enormously from MOC if it is executed correctly, and both can suffer drastically if it is handled inadequately.

Organizational change management: A method of examining proposed changes in the structure or organization of a company (or unit thereof) to determine whether the changes introduce new hazards or increase the risk to employee health and safety, the environment, or the surrounding community.

A wide range of modifications fall into the category of organizational change. It could be as simple as the adjustment of an employee's job duties or the changing of shift schedules for a particular unit of a plant or as complex as the reorganization of a corporate structure. Insufficient planning or incomplete execution of the OCM process can lead to loss of institutional knowledge, deterioration of safety, gaps in responsibilities, and a reduction in personnel competency, all of which could ultimately have devastating results. While OCM is generally perceived as pertaining to operations and maintenance, please be aware that changes to other departments (administration, purchasing, sales, marketing, etc.) could result in indirect impacts to process safety that may not be immediately obvious.

REFERENCES

Canadian Society for Chemical Engineering, *Managing the Health and Safety Impacts of Organizational Change*, Ontario, 2004.

Center for Chemical Process Safety (CCPS), *Guidelines for Technical Management of Chemical Process Safety*, New York, 1989.

Center for Chemical Process Safety (CCPS) Technical Steering Committee, *Retaining Corporate Process Safety Memory*, New York, 2003.

Center for Chemical Process Safety (CCPS), *Guidelines for Risk Based Process Safety*, New York, 2007a.

Center for Chemical Process Safety (CCPS) Technical Steering Committee Workshop, *Organization Management of Change*, New York, 2007b.

Center for Chemical Process Safety (CCPS), *Guidelines for Management of Change for Process Safety*, New York, 2008.

Center for Chemical Process Safety (CCPS), *Guidelines for Acquisition Evaluation and Post Merger Integration*, New York, 2010.

Chemical Manufacturer's Association, *Management of Safety and Health During Organizational Change*, Washington, D.C., 1998. This organization is currently known as the American Chemistry Council.

Contra Costa County, *Management of Change for Organizational Changes*, Section B, Chapter 7, Contra Costa County, CA, 1999.

Contra Costa County, *Management of Change for Organizational Changes*, Section B, Chapter 7, Safety Program Guidance Document, Contra Costa County, CA, 2011.

Entec UK LTD, Contract Research Report 348/2001, *Assessing the Safety of Staffing Arrangements for Process Operations in the Chemical and Allied Industries*, Shropshire, 2001.

Health and Safety Executive Investigation Report, *The Fire at Hickson & Welch LTD*, Castleford, 1994.

2

CORPORATE STANDARD FOR ORGANIZATIONAL CHANGE MANAGEMENT

This chapter will describe the fundamentals of a management system for OCM as well as the specific procedures and risk assessment tools needed. While the framework and procedure are developed for a corporate OCM system, the same program can be applied to smaller units within a company (e.g., division, plant, unit, or individual). A generic workflow process is described and a case study is presented to provide examples of some of the tools and steps of the OCM process.

2.1 CASE STUDY: BP – GRANGEMOUTH, SCOTLAND (2000)

In May to June 2000, three major incidents occurred at the BP petrochemical complex in Grangemouth, Scotland. There were no serious injuries that resulted, but the proximity of these incidents drew wide attention from the public and regulatory agencies. Each incident had the potential for major injuries or fatalities. In response, BP created a task force which undertook a wider review of all operating units and functions across the complex. Lessons learned included many of the elements of PSM as well as those associated with management of organizational changes. The following is a brief summary of each incident.

- The first incident was caused by an electrical fault that led to loss of power and plant shutdown.
- The second incident was the rupture of a steam line that caused disruption to the local community because the piping ran through a tunnel under the public road separating the north and south sides of the plant and required several days to repair.
- The third incident was a fire that occurred in the fluidized catalytic cracking unit (FCCU).

The task force's fundamental process and operations review of each facility concentrated upon the following factors:

- Startup, normal, and emergency procedures
- Competency of operating teams—including procedures for ongoing training and recertification
- Mechanical, instrument, and electrical system integrity
- Control system testing
- Access, lighting, and housekeeping standards
- Existing risk assessment process [hazard and operability (HAZOP), MOC, unit startups, work permitting]
- Recommendations from prior and current incident investigation teams

The task force investigation lasted two months and generated over 800 recommendations. Later, a core team looked at the more generic root causes complex-wide which had wider organizational implications for the company. A total of 10 complex-wide lessons learned were identified, 3 of which include aspects of organizational change management. These are summarized below:

Organization – Significant organizational and personnel changes had occurred at the management, superintendent, supervisor, and operator/technician levels over a number of years. However, new roles and responsibilities had not been formally reviewed against the old roles and responsibilities. In the absence of comprehensive mapping of tasks, certain former tasks were missed and were no longer anyone's formal responsibility.

Initiative overload – Several groups across the complex complained of being faced with a variety of group, business stream (exploration and production; refining; chemicals), and local initiatives. This was most prominent in the Utilities Division, which, in addition to the normal operation of the power station and other utility systems, was supervising the construction of a new Cogen project, implementing infrastructure improvements, while also restructuring. Experienced and knowledgeable staff had great difficulty keeping pace with changes while maintaining the expected quality of work product. There was too little focus on their core activity – running the power plant.

Competency – There was a need to build awareness and competencies in process safety and integrity management within senior leadership and the organization in order to develop meaningful value conversion around cost versus safety. There was a lack of experience in some areas and limited refresher training plans.

As follow-up to these lessons learned, the complex devoted a considerable amount of effort to clarifying roles and responsibilities for the operation and maintenance of every item of equipment within the utilities infrastructure of the complex. An important aspect of these responsibilities has been to ensure that the individuals have the appropriate competencies.

This case study provides insight into how organizational changes may not always be identified as one of the root causes of an accident until these issues are explored within the context of whether there were management systems in place to identify organizational changes that could impact process safety.

2.2 OCM BACKGROUND

Most companies that handle hazardous materials have a MOC program that addresses changes to equipment, facilities, and procedures. Some have included organizational changes in their existing MOC procedures or in a separate procedure. Organizational changes, whether large or small, can have less obvious but equally serious impacts on an organization's process safety performance. In many instances, it may take months or even years to realize the impact on process safety resulting from an organizational change. In some cases, the impact has only been identified after a serious incident; therefore it is important that these changes be reviewed to assess their impact on process safety.

There are no industry standard procedures for OCM; however, companies that have established OCM programs and a number of industry groups have provided guidance regarding how to manage organizational changes. Examples of generic industry OCM procedures, standards, and guidelines are provided in Appendix B. The CCPS *Guidelines for Management of Change for Process Safety*, Appendix A, discusses the safety implications of organizational and staffing changes and provides some examples of these changes, but other requirements for OCM are only briefly addressed. Some companies have tried to implement OCM as part of their equipment and procedure-based MOC programs, but there are significant differences in the review and risks assessment processes, implementation, and monitoring of organizational changes as compared to traditional changes. This chapter will provide some comparisons between MOC and OCM programs. This chapter will also provide examples from an actual organizational change conducted at one company to illustrate many of the steps within the OCM process.

2.3 MANAGEMENT COMMITMENT

Before developing an OCM procedure there needs to be commitment from senior management. For a corporate program, this will require a champion at the corporate level. If the OCM program is being developed at the facility level, it will require commitment from the plant manager. An important aspect of OCM is communication and participation. The individuals affected by the change should be notified as early as possible in the OCM process to ensure their active participation and understanding of the need for the change. Any organizational change can have a significant impact on an individual's morale and focus. The sooner that affected individuals can provide feedback relating to the process, the more time management will have to address these concerns before the change is implemented. Downsizing or divestiture is one exception where communication of the change may not occur until after the change has been authorized for implementation. For these types of organizational changes, the communication part of the OCM process becomes even more important in order to ensure an orderly and safe transition.

The key components of an effective management system as defined in the CCPS MOC Guidelines are shown below:

- Purpose and scope
- Personnel roles and responsibilities
- Tasks and procedures
- Necessary input information
- Anticipated results and work products
- Personnel qualifications and training
- Activity triggers, desired schedule, and deadlines
- Resources and tools needed
- Continuous improvement
- Management review
- Auditing

Additionally, the following issues should also be addressed in an effective management system:

- Policy
- Standards
- Assignment of authority, responsibility, and accountability
- Documentation
- Metrics

A cornerstone of a good OCM program is a well-established process safety competency element. This requirement was addressed briefly in the CCPS *Guidelines for Technical Management of Chemical Process Safety* under the element Enhancement of Process Safety Knowledge and was added as a new separate element Process Safety Competency in the CCPS *Guidelines for Risk Based Process Safety*.

2.4 OCM POLICY

Any management system should start with a clear policy from the highest level of the organization. This policy should state the principles, commitment, and accountability of the organization and commit to proportionate consideration of all organizational changes, large and small, as even those with no apparent connection to safety need to be given due consideration to confirm whether or not they have impacts on safety. For organizational changes, this implies that the policy should originate from the company board or president and then be reaffirmed at the individual plant level.

2.5 OCM WORKFLOW

There are many similarities between the tasks required for a traditional MOC procedure and an OCM procedure. Table 2.1 summarizes the critical task requirements of sample workflows for MOC and OCM to highlight the similarities.

TABLE 2.1 Comparison of MOC and OCM Workflow Tasks

Workflow Task	MOC Procedure	OCM Procedure
Request for a Change	Specifies the purpose, scope, timing, and safety impact of the proposed change to process chemicals, technology, equipment, and facilities	Specifies the purpose, scope, timing, and safety impact of the proposed change to personnel and/or the organization
Approval to Assess Impact of the Change	This approval is typically for resources to design the change and conduct a PHA or risk assessment	This approval is for resources to conduct a risk assessment of the change
Assess Hazards and Risks	Process hazard analysis or risk assessment	Risk assessment
Approval to Install or Develop Plans	Approve resources to design, purchase, and install the change and address recommendations from the PHA or hazard identification	Approve resources to develop plans to implement the change and address recommendations from the risk assessment

TABLE 2.1 Comparison of MOC and OCM Workflow Tasks *(Continued)*

Workflow Task	MOC Procedure	OCM Procedure
Installation/ Planning	Physical installation of changes or redline of procedures	Development of action plans to implement the change based on the risk assessment
Training and Communication of the Change	Communication of the change to affected employees and training for more complex changes	Training to provide process safety expertise to individuals assuming new roles and responsibilities and communicate to affected employees
Authorization to Startup	Pre-startup safety review and authorization to start up the change	Initial action items completed and authorization to implement the change
Monitoring	As required by the specific MOC, such as when testing a new technology or equipment	Monitor the impact of the organizational change by measuring key safety and health performance indicators
Closeout	Verify all action items have been closed out and all documentation updates have been finalized	Verify that all risk assessment action items have been closed out and all documentation updates have been finalized

2.6 OCM PROCEDURE

The next step in developing an OCM management system is to develop an OCM procedure. Since OCM may need to be implemented at every level in the organization, a good starting point might be to develop an overall corporate standard that could then be modified to work at lower levels in the organization, such as individual plant sites or production units. There are many papers that describe OCM procedures for various types of changes. The purpose for a MOC procedure is to ensure that physical changes do not introduce or increase hazards in the process, whereas the purpose for an OCM procedure is to ensure that organizational changes do not adversely impact an organization's ability to identify, evaluate, and control process hazards. The scope of an OCM procedure should address all types of organizational changes that may occur at the corporate, site, or unit level.

It is up to each individual company to determine whether to combine OCM with an already existing MOC program. Some companies may find it simpler to have separate programs for the two issues, but it is certainly possible to just modify the existing MOC program to incorporate these additional types of changes. The important point is that companies address OCM.

2.7 DEFINITION OF ORGANIZATIONAL CHANGE

The OCM procedure should have a clear definition of what constitutes an organizational change. In traditional MOC programs, a simple definition of a change is anything that is not "replacement in kind." For OCM, the key issue is whether the change impacts the organization's ability to implement process safety-critical tasks or to control process hazards; if not, then from a process safety perspective it is not an organizational change. Many different roles within an organization can have an impact on process safety and they all need to be considered. Some typical examples include: operators, technicians, mechanics, supervisors, engineers, and managers. Also keep in mind that contractors may also affect an organization's process safety and should not be overlooked. Organizational change is defined as:

Any change in position or responsibility in an organization or any change to an organizational policy or procedure that affects process safety.

This includes addition and removal of personnel or positions, changes to the duties of an existing position, etc. This may include structural alterations to an organizational chart and changes in reporting relationships or span of control.

Some organizations make a distinction between organizational changes and simpler personnel changes and therefore use different systems, methods, or tools to review and assess the risks associated with these changes. Personnel change can be defined as follows:

The movement of individual personnel into or out of an existing position or new responsibilities requiring new skills and competencies assigned to an existing position.

An example of a personnel change under this definition may be officially assigning a current employee the role of subject matter expert (SME) for electrical classifications at the site. In this case, the personnel change may be reviewed and approved by the individual's supervisor. As with traditional MOC procedures, the level of review and approval should be determined by the type of change or the complexity of the change. The same would apply to an OCM procedure. Refer to Appendix B, Examples B.1 and B.3, for examples of a personnel change procedure.

Figure 2.1 shows an example of a typical work flow for an OCM procedure. The details of each step will be discussed throughout the rest of this chapter.

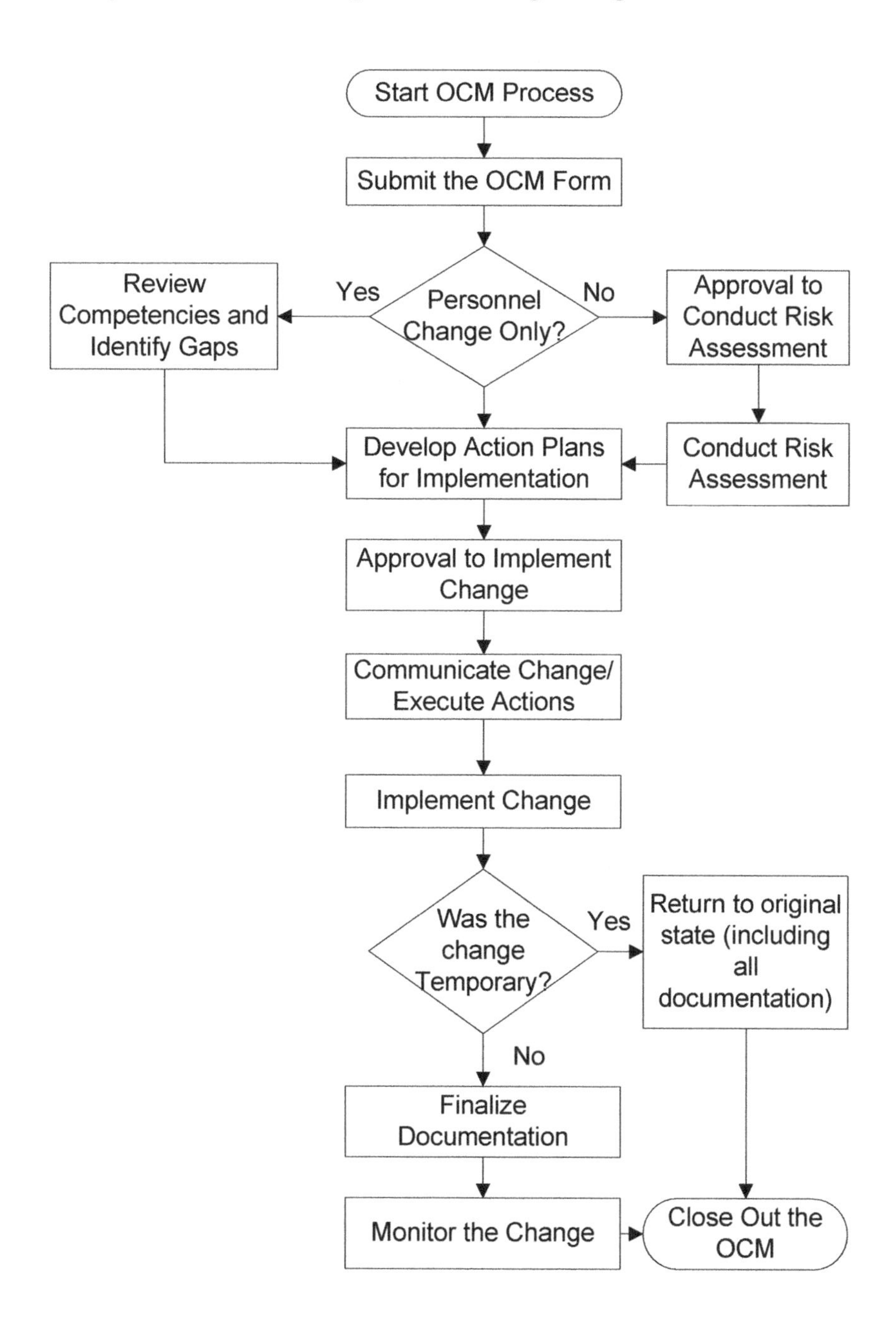

Figure 2.1 Example work flow for OCM procedure.

2.8 ROLES AND RESPONSIBILITIES

The roles and responsibilities of individuals in the OCM procedure need to be clearly defined. The following list includes typical responsibilities within the OCM procedure:

- Determine the job positions and associated duties/ responsibilities that are impacted
- Identify changes covered under the procedure and initiate the OCM process
- Review of the change
- Conduct a risk assessment of the change (team activity led by a facilitator that is trained in the risk assessment methodology)
- Develop an implementation or transition plan
- Authorize implementation of the change
- Assign action items to mitigate identified risks
- Address action items
- Communicate the change and the implementation plan to affected employees
- Monitor impact of the change
- Verify that action items have been addressed and affected documentation has been updated
- Document closeout of the change
- Periodically review and update the OCM procedure

2.9 INITIATE AN ORGANIZATIONAL CHANGE

Once an organizational change is identified, the first step is to initiate an OCM form. Before a change is initiated, some discussion and preliminary review or screening of the proposed change should be completed, similar to scoping out a physical change. Some companies include a challenge step in their OCM procedure to determine whether the change is necessary or desirable before the change is initiated. The organization may limit the individuals who can initiate an organizational change. Typically these individuals would be the supervisors or managers whose personnel would be affected by the change.

The OCM form would include information such as:

- Description of the proposed change, including the positions affected by the change
- Name of individual proposing the change
- Date (when change was initiated and anticipated date of and time criteria for implementing the organizational change)
- Reason for the change
- Duration of the change and transition criteria (e.g., training and certification), if temporary
- Initiator's assessment of potential impact of the change on the organization's ability to control process hazards
- Proposed risk assessment team members
- Plan for implementation of the change

Refer to Appendix B, Example B.2, for sample OCM forms.

2.9.1 Example OCM Case

The XYZ Division of ABC Company decided that they needed to make a significant organizational change. The change involved an overall reduction of 300 personnel (29%), demanning of two installations (reallocation of resources), and a 25% reduction in night shift personnel. They realized that this change would affect the majority of the workforce and could have impacts on process safety. Due to the significant nature of the change, details of the new organization couldn't be widely communicated until the OCM team and company leadership were confident that it could work.

2.10 REVIEW THE CHANGE

The completed form would then be forwarded to the appropriate individual or group for review and approval. Just as with traditional MOC procedures, there may be multiple review and approval steps involved. The initial approval would allocate resources to assess the impact of the change and develop detailed plans to implement the change. The individual or group responsible for providing initial approval of a change can be quite different depending on the extent of the change and how much of the organization would be directly impacted. Organizational changes would typically require approval by a vice president or director at the corporate level, by the plant manager at the facility level, and by the area manager at the unit level. It is important to ensure that the list of approvers includes someone other than the submitter in order to maintain some objectivity in the approval process.

At this point, an implementer would be assigned to take responsibility for implementation of the change. The implementer could be the initiator or someone else who is more experienced with implementing organizational changes. The implementer would assemble a team to review the change and assess the potential risks. The extent of the review and the format would depend on the nature of the change and number of individuals impacted. The subsequent chapters discuss some of the issues which should be reviewed for various types of changes.

2.11 OCM RISK ASSESSMENT

The main component of an OCM procedure is the RA. The OCM risk assessment needs to identify potential risks resulting from implementation of the change as well as any risks associated with the process of implementing the change. "Managing the Health and Safety Impacts of Organizational Change" by the Canadian Society for Chemical Engineering provides a risk assessment checklist to initially assess the safety and health impacts of a proposed change. A worksheet for understanding the organizational change can be found in the "Management of Safety and Health during Organizational Change" by the Chemical Manufacturer's Association, which is now known as the American Chemistry Council. This worksheet is intended to help the "change team" understand and communicate the nature and scope of the pending change. The OCM risk assessment needs to consider the potential process safety impacts under all foreseeable conditions and scenarios, including:

- All activities required to operate and maintain the facility in a safe condition, such as operations, maintenance, and contractor activities
- All activities required for process safety management, such as process hazard analysis, mechanical integrity, incident and near-miss investigations, and management of change
- Effective emergency response

The OCM RA process can be divided into the following steps:

- Preparation for OCM RA which includes:
    - Selecting the OCM RA team
    - Gathering relevant data
    - Selecting the appropriate OCM RA method(s) and tool(s) (Refer to the subsequent chapters in this book

for different types of organizational and personnel changes and specific suggestions for methods and tools most suited to that type of change.)

- Facilitation of the OCM RA using the selected method(s) and tool(s)
- Presentation of OCM RA results to management for review and approval of findings

2.11.1 Preparation – Selecting the OCM RA Team

As is true for a PHA or hazard identification (HAZID), the risk assessment is only as good as the team members (i.e., knowledgeable, skilled, and experienced). The risk assessment team should be led by someone trained in the methods and tools that will be used to identify and assess the hazards. At least one team member should be experienced and/or knowledgeable in the main positions or roles being impacted. At least one team member should be knowledgeable in the process safety management systems most impacted by the change(s). Additional knowledge may be required on the OCM RA team. When selecting a team, the following mix of attributes should be considered:

- Seniority/experience
- Human factors experience
- Understanding of the technical issues (engineering)
- Understanding of the organizational issues (management)
- Mechanical integrity experience
- Health and safety experience
- Risk assessment experience
- Labor/union representation

The team members could include the following roles:

- Department manager
- Engineering
- Mechanical integrity
- Customers (internal)
- Employees
- Contractors
- Co-workers
- EHS manager
- Process safety manager
- Human resource manager
- Union representative

Example OCM Case Study: A team was carefully selected to maximize the opportunity to develop a comprehensive understanding of the change and risk exposure. The members were selected to provide an extensive range of experience within the team, including a detailed understanding of operational needs.

2.11.2 Preparation – Gathering Relevant Data

As with any change, information is required to effectively evaluate the change and assess the risks involved. The OCM form should contain some of the basic information necessary. Depending on the scope and complexity of the proposed change, additional information may be needed, such as:

- Current state and future state organization charts
- Current state and future state of group, department, or organization descriptions, missions, visions, goals, objectives

- Current state and future state of individual and/or position job descriptions, duties, roles, responsibilities, expectations, performance targets, etc.
- Individual résumé(s), competencies, skills, knowledge, training, certifications, qualifications, etc.
- Affected EHS policies, procedures, practices, systems, databases, etc. (identify specific impacts on each)
- Affected engineering, operations, and maintenance procedures, practices, systems, databases (identify specific impacts on each)
- Existing PHAs or hazard identifications conducted of operations impacted by the organizational change

An important set of baseline tools and information to assess how the change will impact the organization and specific job duties are detailed job descriptions [roles, responsibilities, and expectations (RREs)], and competencies (knowledge and skills). Typical duties/responsibilities for affected positions or individuals, if not formally documented, should be developed from the following documents/individuals:

- Job descriptions and task analyses
- Training matrices
- Organization chart
- Duties identified by the manager
- Duties identified by the customers of the affected positions
- Duties identified by the employees and co-workers
- Any safety-related duties including EHS, responsible care, regulatory, and process safety; especially those defined in corporate- or plant-level safety and emergency response procedures

One approach is to maintain a register of individuals and their tasks, roles, and responsibilities related to mitigating major hazards, including contractors who may have a role in process safety. This will eliminate or reduce the level of effort to conduct this exercise as part of the OCM implementation process.

2.11.3 Preparation – Selecting the OCM RA Method(s) and Tool(s)

It is important to understand the OCM RA methods available and the strengths and weaknesses of each. This knowledge will enable the selection of the appropriate method(s) based on the scope and complexity of the change along with the available information from the OCM form and other information collected. Various OCM RA methods and associated tools with their strengths and weaknesses are described in Table 2.2. A more detailed discussion of tools and methods along with multiple examples can be found in Appendix A. It may be helpful to utilize more than one RA method in order to ensure a thorough analysis. Sometimes simple methods are used as a screening tool and then more detailed methods are used as more information becomes available.

TABLE 2.2 OCM RA Methods and Tools Comparison

OCM RA Method / Tool	Pros	Cons
Checklist Review (checklist(s))	Less dependent on OCM RA team experience and knowledge (experience and knowledge is captured within checklist review items)	Checklists should be developed by SMEs to ensure they are complete and effective. Considerable time may go into the development of checklists
	May be complemented with other methods	Highly dependent on thorough checklist. Checklists must be maintained to ensure effectiveness
	May take less time than other methods	Change must match the checklist, if not the checklist may need to be updated or another checklist developed or another method selected
	Generally provides qualitative results	Checklists may be less effective for complex or unusual (unanticipated) changes; checklists are generally not flexible
		Issues that were not anticipated during the checklist development may not be addressed. An updated or new checklist may be needed or selection of another method
		Does not generally provide quantitative results
		Checklists can be very long

OCM RA Method / Tool	Pros	Cons
Brainstorming or High-Level Activity Mapping	Flexible method that can be applied to any type of change	Dependent on OCM RA team experience and knowledge. May be prone to overlook issues
	Doesn't necessarily require extensive background data	Does not generally provide quantitative results
	May take less time than other methods	
	May be complemented with other methods	
What-If or Scenario Review or Simulations (e.g., dry-runs or table top drills)	Flexible method that can handle unanticipated issues	Requires competent facilitator to ensure complete and effective RA
	May be complemented with other methods	Dependent on OCM RA team experience and knowledge and/or effective tools (i.e., detailed job descriptions, competencies, and risk register) (may miss less obvious scenarios or low-frequency events)
	Generally provides qualitative results	Generally takes more time as compared to Checklist Reviews
		Does not generally provide quantitative results

TABLE 2.2 OCM RA Methods and Tools Comparison *(Continued)*

OCM RA Method / Tool	Pros	Cons
Event Tree or Bow Tie	Flexible method that can address unanticipated issues	Requires expert facilitation to ensure complete and effective RA
	Generally used to complement other methods and focus on a few high-consequence issues	Dependent on OCM RA team experience and knowledge and/or effective tools (i.e., detailed job descriptions, competencies, and risk register) and other data that may not be readily available. Additional research or data collection may be needed to support quantitative results
	Can provide quantitative results	Special tools or software may be needed to efficiently capture risk assessment
		Takes considerable time and resources to effectively implement
Task Mapping	Structured review of all activities, roles, and responsibilities	Requires a thorough understanding of current activities, roles, and responsibilities
	Clear delineation of who does what before and after the change	Takes considerable time and resources to effectively implement, especially if the current information is not already available
	Provides good information to use in the implementation plan and communications	Does not generally provide quantitative results
	May be complemented with other methods	

Checklists are commonly used to guide the OCM risk assessment. For organizational changes one company uses the following guidewords which can blend the advantages of a checklist with those of a brainstorming approach:

- Competence
- Communication
- Experience
- Knowledge
- Workload
- Stress
- Priorities
- Staff appraisal
- Authorization level
- Authority
- Work force reaction
- Morale
- Logistics
- Overtime
- Location
- Shift schedule
- Union interaction—if applicable

Another company uses a series of checklists for multiple facets of an organization which may be used to address potential health and safety impacts in specific functions:

- Operations and Safety Effectiveness
- Safety and Health Management
- Safe Work Practices
- Process Safety Management
- Contractor Safety
- Emergency Response

- Safety and Health Regulatory Compliance
- Occupational Health
- Process Unit Operability and Safety Effectiveness
- Craft Safety Effectiveness

For more complex changes, scenario assessments should be conducted. These assessments should focus on whether the new organization can perform the necessary functions in all modes of operation (i.e., normal, startup, shutdown, etc.) and in case of process upsets, incidents, and emergencies. The UK Health and Safety Executive has published a report "Assessing the Safety of Staffing Arrangements for Process Operations in the Chemical and Allied Industries" which contains an example of how to conduct this type of assessment. Scenario assessments would typically be used when the scope of the organizational change affects a large number of individuals with responsibilities for process safety.

As an alternate or addition to scenario assessments, simulated exercises or dry-runs could be developed to test the effectiveness of the new organization. This could be similar to table-top exercises commonly run by companies to test the effectiveness of the emergency response plans as well as the use of process simulators to test response of operations personnel to process upsets. The assessment should verify that individuals can complete their critical safety responsibilities in a reasonable amount of time.

Example OCM Case Study: Desktop scenario testing was undertaken at multiple locations. Desktop scenario testing involved members from a cross section of the workforce. Various foreseeable scenarios were identified for each of the locations. Examples included:

- Power failure
- Failure of critical process

- Emergency response to incident (e.g., chemical release) and follow-up action
- Project implementation (appraise, select through to sanction and execution)
- Corporate requests for data and initiatives

In addition to the desktop scenario testing, process safety and emergency response consultants independently witnessed the emergency response exercises at all facilities. There were no major concerns raised regarding crew numbers or emergency response.

Another risk assessment approach used occasionally is the "Bow Tie" methodology as discussed in the paper "Identifying Key Safety Roles during Organizational Change." The Bow Tie methodology starts with the identification of a potential hazard, such as ignition of a flammable vapor release. On the left-hand side of the diagram are shown the causes of the hazard along with any potential controls or safeguards. On the right-hand side of the diagram is an event tree showing the potential outcomes (i.e., flash fire, flame jet, explosion) and any installed defenses such as fire protection. The approach is to assign an individual role to each control and defense and then determine if any roles are being compromised by the change. In the example provided in this referenced paper, there was a 15% staff reduction at a facility that handles flammable liquids and liquefied gases. The advantage of this methodology is that the impacts can be visualized more clearly though the use of the diagram rather than a table or spreadsheet.

Task mapping is a tedious process whereby all of the affected personnel need to be identified along with the tasks they currently perform. For shift workers, task mapping should include both day and night shifts as well as weekends. Peak workloads and various modes of operation should also be evaluated, such as during major projects, maintenance activities, startup, shutdown, or emergency response. Any special skills or experience necessary to implement these tasks also need to be documented along with the time required to complete each task and any special certifications or qualifications. If some tasks need to be completed simultaneously, this should be noted. Also, any tasks that will be eliminated need to be identified and reviewed to ensure they will not affect process safety. The reassigned tasks will then need to be mapped to the individuals in the new organization. The risk assessment will focus only on those tasks which may have an impact on process safety. Appendix 1 of Managing the Health and Safety Impacts of Organizational Change by the Canadian Society for Chemical Engineering provides a screening checklist for positions that have potentially significant health and safety impacts. Once the relevant duties/responsibilities of the organization undergoing change have been identified, the team needs to identify how each of these functions will be performed after the change. Often the manager of the new organization would be an important resource to identify how these functions will be reassigned.

Example OCM Case Study: Activity mapping was used to evaluate this organizational change. The purpose was two-fold:

- Identify and review activities that would not be transferred into the new organization
- Ensure all required activity is transferred to roles within the new organization

With reduction in personnel from the old to the new organization it was important to ensure that people in the new organization were not overloaded with work. The first part of the activity mapping process required personnel to identify activities that could be removed. Any new roles with excessive hours could be quickly identified with this method and actions taken to address them. Tables 2.3 and 2.4 provide examples of the forms used for this activity mapping exercise.

Following completion of the forms for the new roles, team leaders were able to develop detailed roles and responsibilities for each functional position. This prompted team leaders to discuss and consult with their reports and enabled the workforce to begin to understand how the new organization would function and their role within it.

TABLE 2.3 Activity Mapping Form 1 "Old Role"

Current Role Name:	**Electrical Technician (Two (2) per Shift)**
Date of Mapping:	
Team Members:	
Facilitator:	
Total Time Allocated to Tasks:	

No	Current Tasks	Can task be eliminated? If so, how?	Who will carry out this task in future? (Team/Individual Name)	Approx. Personnel Hours Required / Week for Task	Is the impact of the task on these essential activities high or low?						
					Response to Abnormal Situations	**Maintain Plant Integrity**	**Maintain Plant Availability**	**Manage EHS Systems & Procedures**	**Manage Essential Knowledge & Expertise**	**Is a change involved?**	**Significant Risk**
1	Control and operation of power generation	No	Ops Tech Electrical	3.5	0	0	0	0	0	No	No
2	Assist with startup after production shutdown	No	Ops Tech Electrical	2	0	0	0	0	0	Yes	No
3	Carry out electrical isolations and deisolations	No	Ops Tech Electrical	7	0	0	0	0	0	No	No
4	Carry out electrical risk assessments	No	Ops Tech Electrical	1	0	0	0	0	0	No	No
5	Purchasing and control of electrical stores	No	Ops Tech Electrical	2	0	0	0	0	0	No	No
6	Supervise and work with vendors	No	On-Shore Support & Campaign Tech	2	1	2	0	0	0	Yes	No
7	Complete electrical handover	Yes, now included in operation handover	Ops Tech	3.5	0	0	0	0	0	Yes	No
8	Carry out front line fault finding, maintenance, and repair	No	Ops Tech	36	2	1	1	1	1	Yes	No

"0" – Low; "1" – Medium; "2" – High

TABLE 2.4 Activity Mapping Form 2 "New Role"

Current Role Name:	Ops Tech Electrical (1.5 on Shift)
Date of Mapping:	
Team Members:	
Facilitator:	

***This Assessment**	**Will not be submitted to the operations manager and site manager for endorsement.**	(delete as appropriate)

Total Time Allocated to Tasks:		**Is it possible that the proposed change could cause any of these adverse effects? (Reference Checklist 2) Y/N**

No	Current Tasks	Which role was this task mapped from in old organization?	Approximate personnel hours required per week for task in current organization	Significant risk issues involved?	Excessive workload, fatigue, ill-health	Lack of competence	Poor communications	Deficiencies in team working	Lack of motivation	How are the identified risks controlled or mitigated in the current organization?	What else needs to be done to control these risks?	Will the risks be adequately controlled by these actions?
42	Handover & daily logs	Ops Tech	2	N								
43	Training & assessor	Ops Tech	1.5	N								
44	Cleanups & area tours	Ops Tech	0	N								

TABLE 2.4 Activity Mapping Form 2 "New Role" *(Continued)*

No	Current Tasks	Which role was this task mapped from in old organization?	Approximate personnel hours required per week for task in current organization	Significant risk issues involved?	Excessive workload, fatigue, ill-health	Lack of competence	Poor communications	Deficiencies in team working	Lack of motivation	How are the identified risks controlled or mitigated in the current organization?	What else needs to be done to control these risks?	Will the risks be adequately controlled by these actions?
45	Desk operation duties	Ops Tech	40	Y	1	2	0	0	0	Currently handled by experienced process operators some of whom were already multi skilled technicians in inst and elect trades which have provided a core of competent staff	A core of competent staff is already engaged in daily operations. In addition, an individual training plan has been developed for each person. Further training and coaching are carried out by the core experienced personnel in process control room activities and outside operations. The training plan is in place and training is in progress. The competencies are entered into the competency database	Y

"0" – Low; "1" – Medium; "2" – High

TABLE 2.4 Activity Mapping Form 2 "New Role" (*Continued*)

No	Current Tasks	Which role was this task mapped from in old organization?	Approximate personnel hours required per week for task in current organization	Significant risk issues involved?	Excessive workload, fatigue, ill-health	Lack of competence	Poor communications	Deficiencies in team working	Lack of motivation	How are the identified risks controlled or mitigated in the current organization?	What else needs to be done to control these risks?	Will the risks be adequately controlled by these actions?
46	Reading	Ops Tech	14	N								
47	Taking Samples	Ops Tech	1	Y	1	2	0	0	0	Currently handled by experienced process operators some of whom were already multi skilled technicians in inst and elect trades which have provided a core of competent staff	A core of competent staff is already engaged in daily operations. In addition, an individual training plan has been developed for each person. Further training and coaching is carried out by the core experienced personnel in process control room activities and outside operations. The training plan is in place and training is in progress. The competencies are entered into the competency database	Y

"0" – Low; "1" – Medium; "2" – High

2.11.4 Facilitation of the Risk Assessment

In conducting the OCM risk assessments, it is important to consider previous incidents, upsets, and exercises to determine where significant risks have been identified. Human reliability needs to be considered in the risk assessment, particularly human failures that could result from issues mentioned in the checklists above.

One key consideration is work overload in the new organization. The risk assessment should consider not only the normal workload but also the peak workload and working significant amounts of overtime, such as during turnarounds or projects. Work overload can lead to:

- Omission or poor execution of safety tasks
- Fatigue leading to reduced reliability, errors, poor judgment, or taking shortcuts
- Multitasking that could prevent quick response or adequate execution

The assessment should also consider the following issues:

- Staffing to address previous incidents
- Conflicting priorities
- Emergency, transient, and normal operations

The team then needs to determine and assess the risks posed by the change such as:

- Duties/responsibilities that are not clearly reassigned
- Duties/responsibilities assigned to different individuals who may not be adequately trained
- Loss of organizational knowledge, technical expertise, and corporate memory
- Duties that will be eliminated and not transferred to the new organization

In assessing any deficiencies in reassignment or loss of functions, it is important to assess whether the proposed change will impact the following:

- The ability to identify new or modified regulatory requirements and industry standards
- The ability to assess process hazards, including participation in PHAs and mechanical integrity programs
- The ability to mitigate process hazards, such as having sufficient time to respond to a deviation or an upset condition
- The ability to detect and respond to accidents/incidents, such as employee rounds and emergency response team availability

As part of the assessment, existing PHAs or hazard identifications conducted of operations impacted by the organizational change should be reviewed to determine if any safeguards listed may need to be modified and the risk ranking updated accordingly.

During the OCM RA, the RA team should develop a list of process safety measures to monitor during and after the transition to assure effective process safety performance is maintained. The measures should be relevant to the change being made and readily indicate effective performance or issues. Ideally, there should be some baseline data from before the change for comparison. Process safety measures might include:

- Work group overtime
- Maintenance backlogs
- Maintenance and mechanical integrity quality
- Sickness absence/health records
- Near-misses and process safety events (leading and lagging indicators)

- Average plant availability
- MOC quality
- PHA or hazard identification quality

2.11.5 Documenting the Risk Assessment

Documentation of results for each part of the risk assessment (i.e., each duty/responsibility) could follow the standard spreadsheet format for any PHA or hazard identification (for more details, refer to CCPS *Guidelines for Hazard Evaluation Procedures*):

- Deviation from normal organization
- Cause or the reason for the change
- Consequence or the impact of the change on process safety
- Safeguards are the competencies of the impacted individuals and applicable procedures
- Risk ranking
- Recommendations may include the need for additional quality assurance, procedures, job aids, training and/or temporary or permanent reassignment of responsibilities

Where findings (unacceptable or intolerable risks) are identified, recommendations (appropriate mitigation measures) should be identified. The findings and recommendations including various options available to mitigate the risks should be presented to management for consideration. Upon acceptance of the findings and approval to implement the selected options, action items should be assigned to specific individuals for implementation. Critical path action items should be identified, as well as whether the action item needs to be implemented prior to startup of the change or whether it can be implemented after startup of the change. For recommendations which can't be completed prior to implementation of the change, it may be necessary to identify interim actions until the final recommendations can be fully implemented (e.g., contracting out relief device sizing until a new SME is trained). Typical recommendations might include the following:

- Training to ensure that individuals taking on new duties/responsibilities have the appropriate knowledge and skills. A temporary reassignment of responsibilities may be recommended as part of the training to mentor the individual
- Additional quality assurance steps or procedures to verify critical activities meet performance requirements. The steps or procedures may be temporary or permanent
- Development of new, or changes to, existing procedures and job aids
- Hiring new personnel or temporarily contracting personnel to replace skills being lost
- Adding staff to meet minimum staffing requirements, such as for emergency response or plant startups
- Development of critical process safety competencies (knowledge and skills)

As with any other risk assessment, recommendations that are not chosen for implementation should have the reasons for that decision documented.

2.12 ACTION AND IMPLEMENTATION/TRANSITION PLANS

Once the risk assessment is completed, the findings should be summarized and an action plan developed. This action plan to address risk assessment findings should be part of an overall implementation plan developed for the change. The action plan should clearly identify the critical path for action items, including which action items need to be completed before implementation of the change and which can be addressed after the change is implemented. Confirmation that these preimplementation action items have been addressed should be part of the process to authorize implementation of the change.

One action may be to provide the individuals in the new organization with the skills to implement any new tasks that they may be assigned. Other actions may include new procedures or addition of new automation. Sometimes temporary actions may be necessary, such as scheduled overtime, temporary staffing, or a phased transfer of responsibilities.

The implementation or transition plan should detail how the change from the old to the new organization will occur. Each unique site or unit should consider having its own plan specific to the location. Items to address in the plan may include:

- The critical path for various steps in the implementation or transition plan including prerequisite conditions or triggers for each step
- List of activities that need to occur before the change is implemented
- Description of how day-to-day activities will be managed after the change
- Verification that the transfer of knowledge from individuals leaving the organization has occurred
- Verification that required training of individuals in the new organization has occurred
- Approval for additional resources, both temporary and permanent, that will be needed
- Timeline for completion of the plan
- Communication plan
- OCM closeout

- List of process safety measures to monitor to assure effective performance
- Quality assurance that the organization is meeting process safety expectations

Communication can have a significant impact on how well the change is implemented. The earlier in the process that communication can occur, the more likely it is that the change will be accepted. A webinar presented in January 2011, "Change Management – The People Side of Change," highlighted the impact of change on individuals. Ineffective communication is listed as one of the seven greatest change management obstacles. Any organizational change will cause stress for all individuals in an organization regardless of whether they are directly affected by the change. It is important to deal with the emotions and agendas of individuals. Many individuals will go through a transition curve from initial denial, through resistance and exploration, until they finally accept the change.

A senior member of the organization should give authorization to implement the change. The first step would be to communicate the change to all affected employees. In some cases, due to the sensitive nature of the change, employees who would be best suited to be part of the risk assessment team cannot be informed of the change until it is formally announced. This is similar to an emergency change implemented under traditional MOC programs. For emergency changes, approval for the change should still be given by an authorized individual, but all of the paperwork need not be completed before the change is implemented. In this case, the risk assessment may not be completed with input from these affected individuals. At a minimum, a preliminary risk assessment should be completed and approved. Shortly after the change is announced (e.g., within 30 days), a risk assessment involving all of the relevant individuals needs to be completed to ensure that nothing was missed in the preliminary risk assessment. This approach has an inherent risk in that if the subsequent risk assessment identifies major risks associated with the change, reversal of the change may be difficult if not impossible. For this reason, these preliminary risk assessments without all of the key people should be the exception, and not the norm, for organizational change management.

2.12.1 Example OCM Case

Each of the locations developed a "transition plan" detailing how they would move from the old organization to the new. The purpose of the transition plans was to:

- Describe how the day-to-day activities were to be managed during this transition phase
- Identify those activities to be completed before the new organization could start
- Identify individuals leaving the organization and ensure the transfer of relevant knowledge was completed
- Identify any additional resources required
- Identify timescales for the above activities to take place

The changes within XYZ required a number of personnel to take on additional (or new) roles and responsibilities. In order to ensure personnel were suitably competent for their new roles, detailed competency matrices were developed.

There were many training and familiarization requirements identified for various positions. The following criteria were used to populate the competency matrices.

X	Training and/or competence required
O	Training and/or competence optional
Y	Training and/or competence achieved
X-1	Training and/or competence required prior to transition
X-2	Training and/or competence required within 12 months of transition

2.13 POSTIMPLEMENTATION MONITORING

Regardless of when the risk assessment is completed, it is important to continue to monitor the impact of the change on the organization after the change is implemented. This can be accomplished by monitoring various safety and health indicators that are specific to the change and the potential risks that have been identified. Some examples of leading and lagging indicators which can be monitored include:

- Work group overtime
- Maintenance backlogs
- Maintenance and mechanical integrity quality
- Sickness absence/health records
- Process safety near-misses and process safety incidents
- Average plant availability
- MOC quality
- PHA or hazard identification quality
- Incident investigation quality

Benchmarking these indicators should occur prior to implementation of the change in order to obtain a relevant comparison before and after the change. For more examples of process safety indicators reference the CCPS publications *Process Safety Leading and Lagging Indicators* and *Guidelines for Process Safety Metrics*. The implementation plan should specify how long these metrics need to be monitored (assuming they do not show a significant impact on safety) before the OCM form can be closed.

2.13.1 Example OCM Case

Sets of performance indicators have been identified and are being monitored on a monthly basis. The key performance indicators (KPIs) selected are those that are commonly measured within the business and were felt to provide an indication of whether the new organization is performing better than the old (or at least not getting worse). Some of the KPIs used are listed below:

- Number of safety audits performed
- Number of days away from work cases (DAFWC) recorded
- Number of high-potential near-misses recorded
- Number of imperfect days recorded (day with a process safety event, near-miss, chemical release, etc.)
- Number of sickness days recorded
- Number of overtime hours being worked
- Production efficiency
- Plant process availability
- Maintenance backlog
- Operational excellence costs versus budget
- Budget versus forecast

2.14 CLOSEOUT

The final task is for the implementer to verify that all action items, including any post-startup action items, have been completed and that the affected documentation has been updated.

Closeout of the OCM should include a final review of the implementation plan to ensure all aspects have been completed or addressed. All postimplementation action items need to be addressed. All key performance indicators need to indicate that safety has not been compromised. The final task is to verify that all of the organizational safety information such as organizational charts, individual job descriptions, training matrices, risk registers, required staffing levels, etc., have been updated as required. For temporary changes, all temporary documentation should be removed and the original documents or redlines need to be returned to their original condition. Once all of these activities are completed the OCM can be closed. Some companies allow the implementer of the OCM to close it, while others require a sign-off from a senior person or group.

Although the procedures discussed here are based on experiences that companies have had in managing organizational change, sometimes things don't always work as they were planned. It is important to learn from the experiences of actual organizational changes and address the lessons learned by making changes in the OCM procedure. The final step in any good management system is to review the system periodically through experience and auditing and make changes in the spirit of continuous improvement.

2.15 CONCLUSION

In order to effectively manage organizational change, a company needs to start with a procedure. This can either be a stand-alone procedure or modifications to an existing MOC procedure to incorporate organizational changes. The procedure should identify the types of changes covered by the procedure and should include a thorough risk assessment of the change to ensure that process safety does not suffer due to the change. Various types of organizational change can take place safely as long as they are properly managed.

REFERENCES

Broadribb, M., *Lessons from Grangemouth*, A Case History Presented at the Center for Chemical Process Safety (CCPS) Conference, 2004.

Canadian Society for Chemical Engineering, *Managing the Health and Safety Impacts of Organizational Change*, Ontario, 2004.

Center for Chemical Process Safety (CCPS), *Guidelines for Technical Management of Chemical Process Safety*, New York, 1989.

Center for Chemical Process Safety (CCPS), *Guidelines for Risk Based Process Safety*, New York, 2007a.

Center for Chemical Process Safety (CCPS), *Process Safety Leading and Lagging Indicators*, New York, 2007b.

Center for Chemical Process Safety (CCPS), *Guidelines for Hazard Evaluation Procedures, 3rd Edition,* New York, 2008a.

Center for Chemical Process Safety (CCPS), *Guidelines for the Management of Change for Process Safety*, New York, 2008b.

Center for Chemical Process Safety (CCPS), *Guidelines for Process Safety Metrics*, New York, 2009.

Chemical Manufacturer's Association, *Management of Safety and Health During Organizational Change*, Washington, DC, 1998.

Davidson, P.A. and Mooney, S.D., *Identifying Key Safety Roles During Organizational Change*, Unilever, New York, 2009.

Entec UK LTD, Contract Research Report 348/2001, *Assessing the Safety of Staffing Arrangements for Process Operations in the Chemical and Allied Industries*, Shropshire, 2001.

Harspt, F., *Change Management – The People Side of Change*, Webinar to Center for Chemical Process Safety (CCPS), 2011.

Health and Safety Executive Information Sheet, *Organisational Change and Major Accident Hazards*, CHIS7, Castleford, 2003

Wincek, J.C., *Organizational Change Risk Assessment One Company's Approach*, Croda, Incorporated, Edison, 2009.

3

MODIFICATION OF WORKING CONDITIONS

There are various types of changes that fall under the general category of organizational change. This chapter focuses on planned actions such as modifications to working hours and shift schedules, the opportunity or requirement to work overtime, and relocation of personnel. It also includes a discussion of changes to staffing arrangements in relation to startups, shutdowns, turnarounds, upsets, emergency conditions, and severe weather events during which inadequate manpower can have devastating consequences. To provide some practical insight into the need to responsibly supervise these types of changes, 2 case studies are presented here as examples of what can happen when management of organizational change is disregarded.

3.1 CASE STUDY: ESSO – LONGFORD, VICTORIA, AUSTRALIA (1998)

Through 1991, engineers had worked alongside operators at the Esso gas facility in Longford, Victoria, Australia. Corporate management at Exxon, the owners of Esso Australia, made the decision to transfer all engineers from the facility in Longford to a location in Melbourne as part of a restructuring agenda. This relocation occurred in 1992. The only engineer to remain at the Longford facility was the plant manager. The engineers, who were responsible for initial design, improvements, and monitoring, made periodic visits to the Longford site. They were available by telephone for inquiries from the staff there, although this required initiation by individuals at the site. The intended goals of this move were increased empowerment of the staff and reduced manpower on site.

Within the Esso gas facility, there was a unit dedicated to removing ethane, propane, and other light hydrocarbons from natural gas via absorption into lean oil. This oil, now made rich by the addition of these hydrocarbons, was later distilled to remove the hydrocarbons and recycled back into the system. This distillation process involved the use of a reboiler and a fractionation column. The cold (tube) side of the reboiler contained the rich oil and the warm (shell) side contained lean oil. On September 25, 1998, a process upset caused the lean oil pump to stop, disrupting circulation of the warm lean oil through the reboiler for several hours. As a result, the overall temperature of the reboiler dropped to the temperature of the rich oil, which was −54 °F (−47.8 °C). Ice formed on the outside of the reboiler while plant personnel worked to bring the upset back to normal operating conditions.

The reboiler was not designed to withstand low temperatures or thermal shocks. When operators restarted the flow of lean oil, it caused a brittle fracture in the reboiler resulting in the release of hydrocarbon vapor and liquid. During the explosions and fire that ensued, two employees were killed and eight were injured. Natural gas supplies throughout Victoria were disrupted and did not return to full service for more than two weeks. With no alternative sources, many of the company's domestic and industrial customers were without natural gas for most or all of the time it took for the facility to come back online.

During the investigations that followed, Esso insisted that its operators had been provided with adequate training for their job responsibilities. When operators within the unit were interviewed, it became evident that they had indeed been trained for their positions but did not have a comprehensive understanding of the concepts or reasons behind their actions. Some of the unit supervisors, as well as the facility manager, were unaware of the limitations of the reboiler. According to findings from these investigations, Esso did not follow a systematic procedure to determine what effects the relocation of its Longford engineers may have on process safety. It was expected that engineers would be monitoring operations from Melbourne. Many of the charts from facility data recorders were typically thrown away rather than being sent to Melbourne, and on the day of the incident approximately 30% of the recorders onsite were not functioning due to lack of paper or ink.

3.1.1 Lessons Learned

The relocation of the majority of the Esso engineers to an offsite location meant that operational staff and design staff were no longer in close proximity. Supervisors and operators could phone engineers with specific questions, but they did not have the casual daily interaction with them that existed before the move. This can lead to a situation where people may not even realize they should be asking for help with something, and subsequently problems creep up unexpectedly. Similarly, the engineers did not continue to gain day-to-day insight into practical unit operations as they had when they were an integral part of the plant. Following the relocation, management did not consistently follow up with engineers and operators to ensure regular and effective communication was maintained. Had the engineers remained at the Longford facility, the informal exchange of knowledge between the two groups of employees could have provided each side with valuable information that may well have mitigated, or even prevented, this catastrophe.

3.2 MODIFYING LOCATION, COMMUNICATIONS, OR TIME ALLOCATION FOR PEOPLE

As illustrated in the case study of the Longford explosion, moving personnel to a new location presents its own challenges. When employees from a variety of disciplines work in close proximity, they develop relationships independent of their prescribed job duties. Casual interaction among maintenance staff, engineers, management, and control room employees fosters a sense of camaraderie and a feeling of accountability for each other. With this familiarity comes a more relaxed attitude toward communication. If an operator sees an engineer regularly at work and discusses mundane topics, it becomes comfortable to later approach that engineer with a technical question or a concern about an unusual occurrence within the unit. Issues can be addressed directly and immediately during informal chats throughout the course of the workday.

The separation of staff disciplines should only be considered after a thorough analysis of their interactions has been performed. Transferring a portion of this staff to a different site means that now a telephone or email contact must be made to get questions answered. This may appear to be a minor change in approach, but it transforms a casual conversation during a coffee break into a more formal discussion, requiring a certain degree of interruption of daily activities for both parties. Additionally, this exchange of information may take hours or days when the parties are separated and must rely on phone messages or email instead of face-to-face contact. The issue becomes more pronounced as turnover occurs—there will come a time when none of the existing operators and engineers have ever met each other or worked together. When the physical separation of disciplines is considered, OCM procedures should assess potential impacts with regard to the efficiency of each to perform its duties. Employees should be given the opportunity to express views and present concerns based on their experience working within the existing structure. Documentation of baseline competencies should focus on each discipline and their current functions individually, with special attention being given to the interactions between disciplines and how those interactions will be affected by the move.

3.3 CASE STUDY: CHANGES IN SHIFT SCHEDULES AND STAFFING DURING TURNAROUNDS

An ethylene unit at an unnamed facility was preparing for startup following a planned turnaround. Employees in the unit typically worked 12-hour shifts during startups as opposed to the 8-hour shifts that were scheduled for normal operations. This allowed for the availability of additional staff members to handle the increased number of tasks required during a startup. There was pressure from management, as well as within the unit, to complete the turnaround and resume operations within three weeks. (Note: The name of the company, location, and date for this case study were not specified in the reference.)

For this particular startup, hourly employees refused to work standard 12-hour shifts, though the salaried foremen and managers agreed to the arrangement as usual. Hourly employees, including control room operators, were allowed to work 8-hour shifts instead. This resulted in a shift change of management and supervisory personnel at 7 a.m. and 7 p.m., while operators and other hourly employees had shift changes at 6 a.m., 2 p.m., and 10 p.m. In addition, two professional engineers were added to each 12-hour shift, though their duties were ambiguous.

A simplified diagram of the process is shown in Figure 3.1 for reference. At 2 a.m. on the day of the startup, personnel began the flow of cold liquid into the demethanizer column, which was normally kept at −4 °F (−20 °C). Two hours later, a liquid level should have been recorded within the column, although no one noticed the lack of this information due to distractions caused by other issues. This malfunction was not noticed until 7 a.m., 5 hours after flow was initiated, when the temperature at the top of the column had dropped to −115 °F (−81.7 °C) and the level of reflux drum increased from zero to full in 10 minutes. Employees should have realized that these readings were evidence that the column had flooded, overflowing into the downstream reflux drum and flare knock-out drum, even though the two high-level indicators on the knock-out drum were not registering liquid levels at the time.

No one investigated the cause of these anomalous instrument readings until 12 p.m. that day. A thorough inspection of the unit found that the wiring for the level indicator on the demethanizer column had been disconnected and the knock-out drum was isolated from its level indicators by closed valves. Liquid from the column had now filled the knock-out drum and was entering the flare stack. The low temperature of the liquid caused the stack to become brittle and fail. The leaking liquid from the system did not catch fire, and no one was injured, but this was a significant process safety near miss.

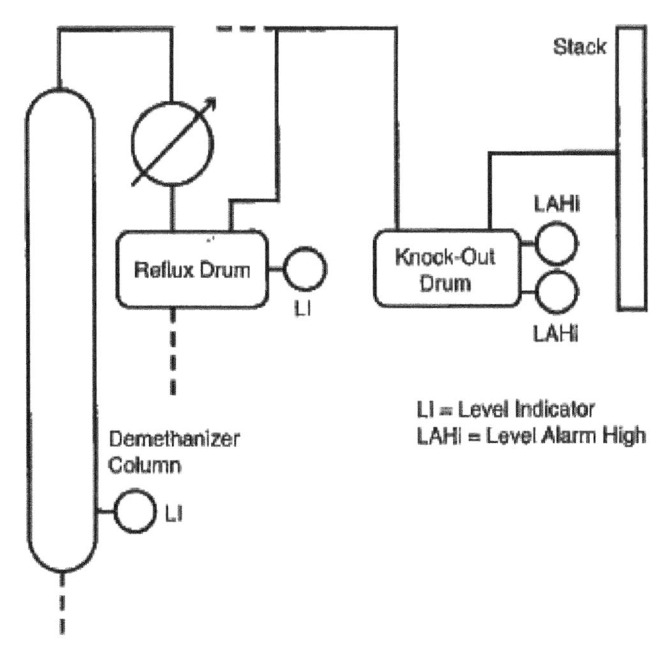

Figure 3.1 A simplified schematic of the demethanizer column and associated equipment.

3.3.1 Lessons Learned

Modifying the working hours of a portion of each normal shift, while keeping the rest of the shift on its normal schedule, meant that the cohesion of each shift was destroyed. While the engineers brought on shift for the startup outranked the shift managers, no one had an understanding of whether the engineers were present to provide advice or to make decisions and instruct the other personnel. It was unclear whether the engineers should remain outside observers or become involved in hands-on operations during startup procedures. All of these factors led to a breakdown in organization at a time when communication within the shift needed to be fluid and transparent. This resulted in confusion over responsibilities and contributed to an overall reduction in the competence of all staff members involved in the startup of the ethylene unit.

3.4 CHANGES TO TERMS AND CONDITIONS OF EMPLOYMENT (E.G., HOURS, SHIFTS, ALLOWABLE OVERTIME)

It is common for shift employees to develop a cohesive team mentality and approach to their duties, much like the members of a soccer team. In an efficient unit, the members of a shift operate as one entity, anticipating each other's moves and unconsciously cooperating on tasks. The defense clears the ball; the midfielders move it down the field and pass to the striker, who takes a shot at the opposing goal. Movements are fluid and somewhat automatic. If changes are made to the structure of a shift without a thorough understanding of how that shift operates as a team, this solidarity can be disrupted. Co-workers may no longer have an inherent understanding of their responsibilities within the group as they relate to each other. Team members may be less confident in the abilities of a replacement forward once one of the starters has been injured or removed from the game. If modifications are only made to some of the shift employees' schedules, this disturbance can escalate. Replace only one forward instead of the entire front line without running a few practice drills using this line-up, and coordination and goal-scoring can suffer. Overlapping of shifts can mean that in the event of an emergency, it is unclear which shift or which employees are responsible for certain duties. This can cause delays in response time, duplication of actions, and a general confusion at a time when efforts need to be streamlined. Making any of these types of changes without first having a clear picture of how employees interact with, and rely on, one another can be detrimental to both efficiency and safety.

An increase in the amount of allowable overtime could occur for any number of reasons. There may be a unit turnaround scheduled in which extra manpower is required for a limited period of time. Internal or external audits could mean that some staff members have to be present for review meetings during off-shift hours. Staffing cutbacks or a hiring freeze may result in a longer-term requirement for fewer personnel to perform the existing job responsibilities. Planned-for temporary staff reductions, as in the case of employees vacationing for the holidays or extended illness, can mean short-term opportunities for additional hours for remaining workers. In any of these circumstances, the opportunity for overtime can lead to cases of fatigue and overwork for employees eager to earn extra money or required to come in during different working hours. OSHA has recognized overtime work hours as a human factor that is frequently overlooked during PHAs. This omission can lead to a situation where management and other decision-makers are not aware of the inherent dangers of unsystematically allowing or requiring employees to work overtime hours. The United States Chemical Safety and Hazard Board (CSB) issued several recommendations following incidents at BP Texas City in 2005. In response to one of these recommendations, the American National Standards Institute (ANSI) and the American Petroleum Institute (API) have developed Recommended Practice 755, which presents guidelines for fatigue prevention in the refinery and petrochemical industries.

Another key consideration when changing people's shifts or hours is the staffing for the emergency response team. In many cases, this team is composed of a variety of employees for whom this duty is in addition to the normal assignments for their work classification. While it is easy to ensure the normal assignments are covered when rearranging shifts and hours for a department, it may be less obvious to ensure that the emergency response team is adequately staffed at all times since the team members come from all over the plant. The employees who are members of the emergency response team should have their work schedule flagged to ensure that changes to their schedule also consider the overall emergency response team staffing as well. Conversely, the employee's home department should be notified when overtime was required for an emergency.

While it is essential to understand how tasks are to be handled in a particular unit in the event of an emergency, and some of the personnel are called upon for emergency response duties, this would not typically be considered an OCM issue. These staffing arrangements should be preplanned and accounted for in the normal job assessments for that unit.

3.5 STAFFING DURING TURNAROUNDS, FACILITY-WIDE EMERGENCIES, OR EXTREME WEATHER EVENTS

Regardless of the location of a facility, there are bound to be occasional weather issues. Blizzards, hurricanes, severe thunderstorms, and tornadoes are some examples of natural phenomena that can have an unpredictable impact on operations. Emergencies within the facility, such as a fire or explosion, can also create extreme confusion and necessitate swift actions to prevent escalation of the emergency. When a unit or an entire facility only maintains the minimum number of employees required to complete job duties at any given time, unexpected upsets such as erratic weather conditions or plant emergencies can wreak havoc on a workforce that is already barely adequate.

As previously mentioned, human factors experts have evaluated operator activities during both normal conditions and nonroutine situations. One study found that operator workload more than doubled during emergency or upset conditions. In addition, it is important to recognize that newer facilities are largely automated and employees may not be exposed to nonroutine conditions as often as they were in older facilities. As a result, personnel have fewer opportunities to gain experience in coping with incidents. While emergencies and upsets can never be predicted, the prudent management team would err on the side of caution. Creating a structure that includes detailed documentation of job duties and the quick availability of additional personnel can reduce the consequences of an upset or eliminate them altogether. Completing a robust OCM evaluation each time changes are made to personnel numbers is imperative to ensure that the availability of adequate manpower during emergency situations is not neglected.

At certain times during the normal operation of a unit, there will be a need for personnel to perform unfamiliar or uncommon job functions. Turnarounds are planned events that typically occur every two to three years. Old equipment or obsolete relief devices may be removed or replaced at this time. New processes may be tested before being implemented into normal operations. Startup and shutdown procedures, as well as any testing or validation of the processes within a unit, are only performed at these times and may have changed due to the work completed during the shutdown. As a result, many employees become involved in procedures they have either never seen or have rarely been exposed to during their tenure. Startups, shutdowns, and turnarounds are periods in which a production unit is at its most vulnerable in terms of process safety because of these concerns.

One solution to issues of familiarity that may be presented during the course of a well-organized OCM is to incorporate process simulations into the regular schedule of training exercises. These simulations can provide employees with worthwhile guidance for how to deal with startups, shutdowns, turnarounds, and an assortment of emergency conditions in an atmosphere conducive to discussion and feedback. Regular drills are also a good learning opportunity for operators and supervisors. Performing drills on procedures for dealing with process upsets will enable operators to increase their familiarity with what steps to take when things unexpectedly go wrong during the course of a normal shift.

3.6 IMPACTS AND ASSOCIATED RISKS

Once baseline competencies and responsibilities have been agreed upon, the next step in assessing impacts is to carefully review each of the types of changes discussed in this chapter. Since many of the changes in this chapter pertain to handling nonroutine situations, a standard gap analysis may not be sufficient. Performing a walkthrough or talk-through of potential atypical scenarios should be considered. Presume the worst possible results of this change and determine what is required to mitigate or eliminate those results. As with most risk assessments, it may be helpful not only to consider the worst-case result but also to look at less significant results which may be more likely to cause a problem. Gaining insight into the possible ramifications of each of these changes will aid in preventing such ramifications from occurring. Completing a walkthrough, in which every team member is involved in the simulation and allowed to participate in the discussions that follow, is the best method for thoroughly examining all of the potential issues associated with an unexpected event. It can be an effective way of including personnel in the mechanics of any prospective changes, as well as an opportunity to obtain valuable insight from them regarding the practical implications of those changes. Table 3.1 shows an example of such a scenario evaluation. Additional information regarding tabletop simulations can be found in the appendices of this book.

TABLE 3.1 A Potential Atypical Scenario and Mitigation Options

Scenario Description	Concerns	Action Recommended
Flange leak of toxic gas at night. Shift numbers are typically lower at night, but within minimum staffing requirement. One operator works the night shift in the unit in question.	It is difficult for an operator to keep track of process conditions during upset or emergency conditions, as they have to personally detect a toxic gas leak because there is only one person who would be in the control room, out in the plant, or loading tankers.	Consider implementation of a plan in which the operator would ask for assistance from an operator in the utility plant if available or to call in the next operator due in to assist with emergency situations. Determine the elapsed time necessary before an adjacent or off-duty operator could report to the scene. Consider alternate routes to the unit if the main route is downwind of such a leak.
	Although there is an alarm in place to alert security if the operator is incapacitated, there is currently no contact between the operator and other personnel onsite except during daytime hours.	Assess the benefits of introducing interaction with other parts of the site during the night shift. Consider regular security checks throughout the night shift for currently undermanned units.

Personnel questionnaires and diaries can be useful here as well. Completing these tools based on the existing structure of the unit, and then reevaluating the information based upon the anticipated changes, can provide for a straightforward comparison. Running a simulation as if the potential changes were in place can help identify exact points in new processes or procedures where further explanation may be needed or where additional modifications may be required to avert disaster.

A number of predictable issues that are inherent when changing an existing organization should be considered during the assessment phase of an OCM evaluation. An initial upheaval in standard activities is to be expected as employees acclimate themselves to new shift schedules, different working hours, or a new job site. This adjustment period lends itself to increased stress and uncertainty about job responsibilities. Breakdowns can occur in the normal avenues of communication between departments, teams, or shifts. A loss of concentration is likely as co-workers discuss the changes or help each other to adjust to new surroundings. Changes in the structure of a shift or meeting new co-workers as a result of location changes mean that it takes time for a team atmosphere of trust and cooperation to develop. This can lower the efficiency of a shift or department until the adjustment is completed. A brainstorming or "What-If" approach may be helpful with bringing out some of these issues. Additional resources regarding these tools can be found in Appendix A of this book.

During emergencies it is difficult to predict human reactions. As a result of the initial chaos that can occur when something unexpectedly goes wrong, employees may become confused about their responsibilities. Experienced personnel may default to relying on memory to complete their typical job duties. One study determined that there can be a reduction in working memory during times of stress, thus impairing an employee's ability to perform his or her job at the very time when efficiency is needed most. A loss of focus can mean that an operator or other specialist is sidetracked from important duties only they can perform and spends valuable time contacting management or emergency personnel. Newer employees or those with less expertise may be left with no clear understanding of how or where they can effectively assist in bringing the situation to a safe resolution. Personnel may focus strictly on solving the issue at hand and forget about routine duties or monitoring of other equipment, thus potentially compounding the existing problem. Worker fatigue may play a factor in how efficiently employees complete their duties or how clearly they can evaluate potential solutions. Example B.5 in Appendix B provides a good method for evaluating the minimum staffing necessary to get a process unit into a safe state in the event of an emergency or upset condition.

The unusual working conditions that occur during a turnaround can also have detrimental effects. Personnel are required to perform tasks that might be completed only once every few years. Even employees who have been on-site for many years will not be completely comfortable performing these unfamiliar and potentially detailed or complex duties. As with emergency situations, the additional concentration required during a turnaround may mean that less attention is given to normal tasks or monitoring duties, with potentially dangerous results. With these unfamiliar duties comes the question of which employees are responsible for tasks that fall outside of normal job descriptions. The introduction of new processes or addition of new equipment during the course of a turnaround exacerbates this problem. Longer working hours or additional shifts may be required to complete the turnaround of a unit in a timely manner. As with emergencies, worker fatigue is likely to play a role in a loss of efficiency or lack of attention to detail during a turnaround.

3.7 SPECIAL TRAINING REQUIREMENTS

A number of steps may be taken to increase the awareness of the impact of organizational change under any of the circumstances described in this chapter. They should become an integral part of the change process, and those responsible for overseeing the change should be skilled in employing these practices. In the case of changes to working conditions, such practices include, but are not limited to, the following:

- **Emergency simulations** – Process simulations can be developed into a regular part of the training program within a unit or facility-wide. These exercises may include procedural questionnaires and job diaries to facilitate a better understanding of the complex interactions of team personnel and to draw attention to quality control concerns with regard to process documentation.

- **Overtime policy** – Establishing and enacting a clear policy for overtime, including how much is allowed and under what conditions, can be fundamental in battling worker fatigue. By reducing the chance of stress, tunnel vision, or lack of focus due to worker fatigue, emergency situations can be resolved more efficiently or perhaps avoided altogether. It is important to note that if the workforce is unionized, adjustments to its overtime policy will likely require changes to the existing labor agreement. Guidelines for determining the potential for worker fatigue can be found in ANSI/API RP 755.
- **Regular teleconferences** – When a sector of personnel, whether it be several employees or an entire job function, is relocated, management must quickly develop a means of replacing their original pathways of communication with other team members or departments. Beginning a habit of weekly teleconferences regarding practical operational or safety-related topics allows for an open forum for discussion of equipment issues, questions about new procedures, and sharing of other valuable information across work disciplines that might not otherwise occur within the new organizational structure.

3.8 CONCLUSION

Changes to working conditions can take a variety of forms. Communication and training are key concerns associated with these types of changes. When teams of people are disrupted by changes to hours or location, communication pathways can change resulting in disruptive consequences. Both formal and informal communication between people must be encouraged and managed during such changes. Training may also need to be reviewed to ensure that these types of changes do not leave someone without adequate understanding and/or resources in an emergency.

REFERENCES

American National Standards Institute/American Petroleum Institute Recommended Practice 755, *Fatigue Risk Management Systems for Personnel in the Refining and Petrochemical Industries, 1st Edition*, Washington, DC, April 2010.

CSB Report, *BP Texas City Investigation Report Refinery Explosion and Fire*, March 2007.

Entec UK LTD, Contract Research Report 348/2001, *Assessing the Safety of Staffing Arrangements for Process Operations in the Chemical and Allied Industries*, Shropshire, 2001.

Kletz, T., *Still Going Wrong! Case Histories of Process Plant Disasters and How They Could Have Been Avoided,* Gulf Professional Publishing, Houston, 2003.

Hopkins, A., *Lessons from Longford: The Esso Gas Plant Explosion*, CCH Australia Limited, Sydney, 2000.

Perron, M.J. and Friedlander, R.H., Process Safety Progress, *The Effects of Downsizing on Safety in the CPI/HPI,* Marblehead, Spring 1996.

4

PERSONNEL CHANGES

This chapter addresses organizational changes involving personnel changes. One of the most common type of personnel change involves downsizing and layoffs. However, there are many other types of change that should also be considered in this category, including retirements, replacement of key plant personnel such as the plant manager or a subject matter expert, replacement of capital project personnel, and strike coverage. In this chapter, you will be presented with two case studies in which personnel turnover issues were not managed appropriately and led to disastrous safety consequences.

4.1 CASE STUDY: UNION CARBIDE – BHOPAL, INDIA (1984)

Many people are familiar with the Bhopal disaster in December 1984, and as is usually the case, there were multiple problems and failures leading up to the catastrophic release of methyl isocyanate (MIC). For this case study, we will focus on the organizational issues related to personnel turnover to provide an example of how such changes can contribute to process safety incidents.

The initial Indian managerial and supervisory staff for the Bhopal methyl isocyanate production unit was trained in Union Carbide's West Virginia plant. As prospects declined and the future of the Bhopal facility became doubtful, the trained staff began leaving for more attractive jobs and were replaced by employees who had *not* received prior training in operating an MIC plant. Additionally, low production volumes seemed to justify workforce reductions. In the methyl isocyanate unit, the workforce was reduced from 3 supervisors and 12 workers per shift to just 1 supervisor and 6 workers.

In the late fall of 1984, plant operations were focused on using up existing stocks of chemicals to prepare for the sale of the plant. However, plans for making the last batches of product were delayed, due to rioting in the wake of the assassination of Prime Minister Indira Gandhi. City authorities imposed a curfew for several weeks which resulted in difficulties with getting the second and third shifts of workers into and out of the plant.

On the evening of December 2, 1984, the second-shift supervisor ordered workers to perform a periodic washing of pipes in the MIC storage area to control corrosion. The worker performing the washing noticed a couple of problems – the slip blind in the pipe to make sure that water did not back up into the storage tanks was missing, and one or two of the bleeder valves at the bottom of the pipes where wash water should have come out were blocked. The worker reported these problems to his immediate supervisor, an operations supervisor rather than a maintenance supervisor, and was told to continue with the washing operation.

Water entered one of the storage tanks, resulting in a pressure rise. Due to a variety of equipment failures, a large cloud of gas escaped from the vent stack. Residents of nearby areas began smelling the gas. Due to a lack of information about what to do, the local residents panicked and began to flee. Medical staff converged on their hospitals and clinics as they heard about the situation, but the initial efforts to treat patients were hampered by a lack of information about the gas or antidotes. Estimates of the number of immediate fatalities caused by the Bhopal gas cloud vary from the official Indian government figure of approximately 2,000 to the 10,000 favored by local activists.

4.1.1 Lessons Learned

When an experienced employee leaves the company, it is crucial to understand all of the roles he or she plays and ensure that these are adequately covered by remaining employees or his or her replacement. When considering the roles and responsibilities of an individual, be sure to include any unique skills, knowledge, and relationships he or she may have but which are not explicitly included in the basic job description. (Is he or she the only person who truly understands the idiosyncrasies of a particular piece of equipment? Does he or she know where to obtain replacement parts which are not readily available? Is he or she an instrumentation wizard whose expertise was just a bonus and not actually part of the job requirements?)

Although people may not want to admit it, having less experienced personnel working in an increasingly sophisticated, computerized manufacturing facility will increase the risk of serious and costly mistakes. Many inexperienced people tend to rely too heavily on information provided by the computer system and may not understand the value of taking a look around in the field. In some cases, employees may not even understand all of the actions that the computer control system is taking or where the devices are located in the field. Even when errors are not caused by inexperience, diagnosing and fixing a problem often takes longer when veteran employees are no longer around to assist. The experienced individuals provide a wealth of knowledge gained from having seen many upset conditions and can provide extensive informal training to newer employees. Without the benefit of experience, additional time and money should be allocated to training in order to bridge this knowledge gap.

When considering downsizing, especially as it pertains to the actual operating staff, a comprehensive evaluation should be made to ensure there is no critical reduction in process knowledge or response times during emergency situations. The temptation to simply demand fixed percentage cuts from all departments must be avoided. The modern control room is a complex sociotechnical system in which a number of factors influence the effectiveness of the operation, and thereby the safety of the facility. A baseline study should be performed, as well as a review of the proposed change, to ensure that the new organizational structure does not reduce emergency capabilities below tolerable limits or introduce significant new hazards. This review should be done by a team of people including control room operators (experienced and inexperienced), staff who would assist during incidents, and management or engineering staff with knowledge of operating procedures, control system configuration, process behavior, equipment and system reliability, and safety. This type of review is essential for bringing staffing issues into the open and making it plain which factors have a bearing on process safety. Example B.5 in Appendix B is a procedure for assessing minimum staffing levels which provides a good template for conducting such a study.

When the number of personnel is reduced due to downsizing or inability to fill a position, be on the lookout for issues related to overloading and fatigue. If the amount of overtime increases significantly due to the change, this could result in operator errors, tasks not being completed in a timely manner, or a decrease in morale. Overload and fatigue can also lead to increases in minor injuries, near-misses, and/or increases in off-specification product. The minimum staffing levels should be evaluated not just for routine activities, but also for safely handling upset conditions or emergency situations which will often require additional activities and attention. Another consideration is possible changes in response times. If process safeguards are premised on a minimum detection time, response time, or repair time which can no longer be met with reduced staffing levels, then consideration should be given to additional automation or alternative safeguards to maintain acceptable risk levels.

Another not-so-obvious impact of downsizing is a general loss of current knowledge. Given that the expectations for process safety are performance based and premised on achieving the current best practices, it is essential that key people tasked with implementing process safety are provided with time and opportunities to understand the latest trends in the field. As people's time is pushed through expanded job expectations due to downsizing, this task of training, researching, and collaborating to keep your knowledge base current can slip. Therefore, companies may end up falling behind in terms of process safety expectations without ever realizing it.

During any organizational change, but especially for downsizing, it is important to not only consider the implications of the revised organization, but to also think about the potential safety impacts during the transition. Whenever possible, phase in the changes to prevent a loss of control through overcomplexity and to minimize peaks in workload. In addition to training, it is important to ensure that ample support and/or supervision by competent people is available for anyone with new safety-sensitive work. Do not hurry through such a change before all necessary training is complete and any new safety measures are in place and functioning. There should be clear criteria available with regard to competency levels which should be verified before the organizational change takes effect.

4.2 CASE STUDY: BAYER CROPSCIENCE, LLC – INSTITUTE, WEST VIRGINIA, USA (2008)

On the evening of August 28, 2008, there was an explosion in the Methomyl unit at the Bayer CropScience, LLC facility in Institute, West Virginia. Two people were killed and eight people were injured as a result of this incident. While there were several root causes of the explosion, this case study is going to look at the contribution of an organizational change related to the capital project being implemented just prior to restarting the Methomyl unit.

The capital project involved the replacement of the Residue Treater with a new stainless steel vessel, which was identical to the old unit other than the material of construction, and the installation of a new distributed control system (DCS) to control the operation of the Methomyl unit. The project liaison from manufacturing was a technical advisor who had no experience in the Methomyl unit. He did have DCS training, but it was regarding a different brand of control system than what was being installed for this project.

On the day of the incident, the Methomyl unit was being started up after an extended turnaround, which included the implementation of this capital project. The flasher bottoms were inadvertently fed to an empty Residue Treater Vessel. In addition, the interlock for minimum temperature was bypassed. Due to upstream problems, the concentration of methomyl in the feed stream and the the Residue Treater were much higher than they were supposed to be. The methomyl in the Residue Treater began to decompose much more rapidly than designed. This led to a runaway reaction and eventual over pressure of the Residue Treater Vessel.

4.2.1 Lessons Learned

The technical advisor assigned to the capital project team was responsible for providing operational input to the project to ensure that the project would be able to meet the requirements for operating the facility in a safe and efficient manner. Unfortunately, since this person did not understand the operation of the Methomyl unit, there were many oversights in the functionality of the control system which contributed to the inability of the operators to safely restart the unit in August of 2008.

The units of measurement for some of the process variables on the DCS displays were different from what the operators were used to (e.g., lbs of product in a vessel instead of % full). This caused confusion in the control room and required the operators to create temporary conversion notes to place next to the DCS monitors. Not all of these conversion notes had been developed at the time of the incident and the use of such conversions could have led to delays in comprehending the information being presented.

Due to limitations of the process equipment, instruments, and piping, the operators had become used to deviating from the standard operating procedures (SOPs) for certain tasks. In one instance, the level control on the Residue Treater was normally put in manual mode and the operators would periodically pump out the liquid using a high flow rate to avoid plugging lines, which frequently resulted when the level control was set to "auto" mode and allowed to pump out at a continuous, slow rate. In another instance, the operators had trouble preheating the clean solvent in the Residue Treater to the minimum operating temperature with the equipment available, so they usually bypassed the minimum temperature interlock on startup so that the decomposition reaction of the methomyl in the feed could complete the solvent heat-up. None of these modifications were reflected in the SOPs, nor were they accommodated in the control scheme of the new DCS. The liaison assigned to the project team from production should have been knowledgeable of the unit operations and these types of deviations, and then modifications could have been made to the control system logic such that the operators would not have to take manual control, which can open the door for process upsets and safety problems. As with any other job position, the knowledge requirements should be fully understood prior to assigning someone to a role on a capital project, and if that person lacks some essential knowledge, then some sort of training program or other accommodation should be implemented.

Frequently process safety is only one of several, sometimes competing, priorities for capital projects. However, any shortcuts made in the initial installation of a project can have lasting implications for the facility. The priorities and objectives with regard to process safety should be clearly outlined as part of the project, and these requirements should be clearly understood by the individuals on the project team.

The potential risks associated with a change will depend on the type and complexity of the process that is being implemented or modified. For many processes, a thorough understanding of the technology or process chemistry is essential for designing and implementing appropriate safeguards for process safety. If appropriate equipment and instrumentation are not specified and installed up front, it is sometimes too costly, if not impossible, to modify after installation. If the project manager or project engineer does not fully understand the potential safety implications of a process, they may not be willing or able to make a reasoned case for the funding necessary to meet process safety objectives or they may unknowingly be talked into taking shortcuts which have adverse impacts on long-term safety in the interest of other priorities.

When documenting baseline competencies for positions on project teams, the approver should be sure to include safety training and knowledge. A key component of this safety knowledge is inherently safer design strategies which are best addressed at the project design stage. There are many industry standards which pertain to project design. Familiarity with these should be included in the baseline competencies. Do not forget to include experience and participation in teams which impact process safety such as PHAs or hazard identification, incident investigations, and pre-startup safety reviews.

The authorizations for changes to a capital project team are a little unique because capital project teams are formed and dissolved along with the project life cycle and personnel assigned to work on a project often changing between initial conception of the project and execution of the project. These changes to the composition of the project team can compound the issues discussed above regarding knowledge and competencies if the baseline requirements are not well defined and managed during each change. Once a project is approved, it may be prudent to issue an OCM for the establishment of the project team with key people assigned and their knowledge and competencies defined. Additional OCMs may also be necessary if changes are needed regarding the key project personnel throughout the course of the project. In some cases, these assignments may be for a full-time role on the project, but in many cases, participation in the project team is only a portion of someone's job—at least for some phases of the project. However, ensuring that people with the right knowledge have the necessary time and authorization to participate in the design and implementation of the capital project is critical. An OCM is a good way to convey the importance of such a key project role throughout the organization.

The actual authorization for a change to (or establishment of) a capital project team will depend heavily on the scope of the project. For a project which is being managed with the plant site personnel, the authorization can be done by the key stakeholders for the project, which typically include, but are not limited to, operations, safety, project engineering, and maintenance. If the project is being managed by a corporate project engineering department, then the approvals may include some of the same plant-level positions in addition to someone from corporate project engineering and safety.

4.3 CHANGES IN PLANT MANAGEMENT (e.g., PLANT MANAGER OR EHS MANAGER)

These top-level employees set the tone for process safety at the plant level. They must not only "talk the talk," but also "walk the walk." Frequently, the people in these roles take on new assignments with specific objectives for the site which may or may not be related to process safety. However, if they fail to provide a strong focus on process safety, the underpinnings of the entire program could be in jeopardy.

When evaluating the baseline competencies for the position, be sure that process safety is included. Consider not only the actual tasks performed by this individual with regard to process safety, but also the vision for process safety at the site.

Since these are very high-level positions within a plant site, these changes will often be authorized by someone at a more senior level or even a corporate role. To facilitate a smooth transition, a plan should be developed with action items at the local level. These action items should be communicated and executed as quickly as possible after the organizational change is communicated. A transfer of knowledge from the prior position holder to the new one is very important so that the new person understands the critical issues with regard to process safety at the site, since some issues may require constant vigilance to avoid a catastrophe.

4.4 REPLACEMENT OF A SUBJECT MATTER EXPERT

There are many examples of subject matter experts (SMEs) at most facilities, such as process expert, emergency relief sizing, risk assessment, mechanical integrity, incident investigation, and PHA leadership. In some cases these tasks are clearly spelled out as part of the job description for these positions, but in other cases people may have developed into de facto SMEs over time. These skills might not be captured as critical competencies for their positions.

When you are replacing a SME, this competency is frequently already included in the job description, making the gap analysis fairly straightforward. In many instances, this SME role may not have a reporting relationship to the safety department although the role is critical to process safety. In these instances, the change should be authorized by the immediate supervisor with input from someone in the safety department responsible for that subject matter.

4.5 REPLACING THE INCUMBENT IN A POSITION THAT DIRECTLY AFFECTS PROCESS SAFETY

This issue applies to both corporate and site employees. It is essential that all process safety responsibilities and activities are captured in job descriptions for these positions, and that incumbents have the same capabilities or there are plans put in place to supplement the skills for the incumbents, to ensure that process safety is adequately covered.

Some companies treat this sort of change as a personnel change instead of an organizational change, the difference being that the personnel change is subjected to a simpler risk assessment and approval process. Refer to the example procedures B.1 – B.3 in Appendix B for examples of the differences between personnel and organizational changes.

4.6 STRIKES, WORK STOPPAGES, SLOWDOWNS, AND OTHER WORKFORCE ACTIONS

Contingency planning for workforce actions should include careful consideration of process safety implications. These types of changes are temporary changes which often do not have specified end dates. When approving a temporary change like this, a reasonable time frame should initially be selected for the change. The situation should be periodically reevaluated and modifications made as necessary. Depending on the extent and duration of a change like this, it might be necessary to authorize multiple temporary changes until the situation is fully resolved. Do not overlook the fact that an additional organizational management of change may be necessary to return to the normal workforce if they have been off the job for a substantial period of time.

Although workforce interruptions like this might happen with little to no notice, management should ensure that decisions and changes which could jeopardize process safety are not rushed. In some ways, this could be compared to an emergency management of change for a physical piece of equipment in which certain key people are consulted and authorize a temporary change until there is time to put together a more thorough management of change package.

Since this type of change affects an entire plant site, even if indirectly, the change should be authorized by the plant manager as well as the senior EHS person at the site. Do not overlook the fact that additional organizational changes may need to be authorized within other parts of the company if people need to be "borrowed" from other locations or roles to back fill for the missing workforce.

Sometimes changes can result in unclear reporting lines. This is especially true for strikes or other workforce interruptions. Updated, accurate organization charts can help, but it may also be necessary to specifically discuss and document these issues up front when announcing the change. In situations like these, people often end up temporarily covering multiple concurrent roles and it may be necessary to clarify back-ups or alternates for key positions to guarantee resources are available when they are needed. Individuals should be empowered and encouraged to discuss and question conflicting instructions or unreasonable demands.

As mentioned in the case of downsizing, companies should watch out for possible problems with fatigue, overwork, and low morale. Workforce interruptions involve significant stressors, which can exacerbate these issues. These potential problems should be included in the initial management of change with appropriate action items, and the situation should be periodically monitored in case additional actions become necessary.

Although training requirements as outlined in the PSM standard are a good starting point, additional considerations should be made – such as adequate communications, clear delegation of authority and responsibility as it pertains to emergencies, and understanding of new, and possibly overlapping, roles and responsibilities.

4.7 EMERGENCY RESPONSE TEAM STAFFING

If an organization has an emergency response team, then consideration for the staffing of this team should be included as part of any personnel turnover, reassignment, holiday season, or staff reduction. It is easy to overlook the fact that organizational changes might result in gaps in coverage for critical skills within the emergency response team. Membership in the emergency response team is often a voluntary task supplementary to a person's main job function. Participation may not be a necessary part of the employee's job description; however, it is critical that there is adequate staffing for the emergency response team during all plant shifts. In some organizations, the emergency response team provides coverage for other facilities through a mutual aid agreement. In these situations, all stakeholders should be notified of and/or involved with reviewing and approving organizational changes which could impact the emergency response team.

4.8 IMPACTS/ASSOCIATED RISKS

The impact of personnel turnover can vary significantly depending on the level within the organization and the number of changes being made. If the change is just for a single person moving into or out of an existing position, then the change may be covered with existing processes for personnel replacement as noted in the next section.

If the change involves the replacement of more than just a single person or position, a team approach should be used to evaluate the impacts and conduct a gap analysis of the change. The team should be comprised of people from multiple functions within the organization, similar to the composition of the PHA or hazard identification team. Decisions regarding who should be included in the evaluation team should take into consideration experience within the affected departments, knowledge of human factors issues, technical capability, organizational and business expertise, and an understanding of risk management concepts. Checklists or "What-If" methods are well suited for evaluating this type of organizational change. Refer to Appendix A for examples of various checklists which may be helpful in evaluating such changes. Additionally, tabletop simulation exercises can be helpful. Refer to Example B.5 in Appendix B for an example of such exercises to ensure a minimum staffing level in a process unit.

The risks presented by personnel turnover can be significant. Many of these issues have already been mentioned in this chapter. Table 4.1 includes several issues that should be considered with these types of organizational changes.

TABLE 4.1 Potential Risks Associated with Personnel Turnover

Loss of process knowledge	Loss of expertise in key process safety areas
Stress	Overload and fatigue
Loss of focus	Inefficiencies due to new roles and responsibilities
Lapses in communication	Confusion regarding responsibilities and reporting relationships
Loss of cooperation	Ineffective capital project installation

4.9 ORGANIZATIONAL CHANGE PROCEDURES VERSUS OCM FOR NEW HIRES, PROMOTIONS, ETC.

All companies have some type of procedure, usually directed by the human resources function, to handle routine new hires and promotions. This management of organizational change procedure is intended not to replace that process, but instead to complement it. In many instances, hiring and promotions do not impact process safety at the facility and therefore do not need to follow this management of change procedure. However, someone familiar with the change and with process safety at the facility should be consulted to determine if this additional procedure is needed for each particular change. If the new hire or promotion *does* have an impact on process safety, then this management of change process should be added to whatever procedures are normally utilized by human resources. Whenever possible, the incumbent person in the position should be included in the assessment of the change to make certain that nothing is overlooked. Refer to Appendix B for an example of a procedure for personnel change. In such cases, the risk assessment may be simplified as compared to assessments for more complex organizational changes.

4.10 CONCLUSION

Personnel changes can encompass simple activities such as hiring a new engineer or complex changes such as operating a process plant during a union work stoppage. Each of these types of changes needs to be reviewed for its potential impact on process safety, if any. Simple changes may be handled with a streamlined procedure in which the manager documents the change, updates the necessary documents, and ensures that essential training and communication take place. More complex changes should be handled using the OCM procedure, including a risk assessment, to ensure that all necessary actions are taken to prevent process safety problems as a result of the change.

REFERENCES

CSB Report, *Investigation Report, Pesticide Chemical Runaway Reaction Pressure Vessel Explosion*, August 28, 2008.

Entec UK LTD, Contract Research Report 348/2001, *Assessing the Safety of Staffing Arrangements for Process Operations in the Chemical and Allied Industries*, Shropshire, 2001.

Petersen, M.J., *Bhopal Plant Disaster - Situation Summary (Draft)*, International Dimension of Ethics Education in Science and Engineering Case Study, March 2009.

5

TASK ALLOCATION CHANGES

This chapter deals with reallocation of tasks to different individuals or the elimination of tasks that may have an impact on process safety. This can occur for a variety of reasons of which some of the more typical causes will be discussed later. The potential repercussions of such changes are examined in the case studies in Sections 5.4 and 5.8.

5.1 DOWNSIZING EXAMPLES

Some years ago, a survey of Swedish chemical process companies belonging to the Swedish Association for Process Safety was conducted regarding the subject of downsizing. All member companies were asked about their experiences with downsizing. Examples of some of their responses are presented in Table 5.1.

Common themes included:

- Loss of competencies
- Neglected activities (e.g., maintenance)
- Difficulty integrating large projects into the plant

TABLE 5.1 Examples of Disturbances, Accidents, and Major Costs Where Downsizing Has Played a Major Role (Sweden)

Respondent	Incident/Performance History	Reasons for Degraded Safety Performance
Member 1	Series of major process safety incidents, environmental releases, and plant shutdowns in petrochemical plant 1	Early retirement of competent personnel, major investment project, neglected maintenance
Member 2	Series of major process safety incidents, environmental releases, and plant shutdowns in petrochemical plant 2	Big investment projects, administrative projects, loss of competent personnel to early retirement
Member 3	Series of major process safety incidents, environmental releases, and plant shutdowns in petrochemical plant 3	Decreased maintenance, outsourcing, loss of competence
Member 4	Series of major process safety incidents, environmental releases, and plant shutdowns in petrochemical plant 4	Neglected maintenance, loss of competent personnel to early retirement, major investment project
Member 5	Lack of competency requiring rehiring of personnel	Amalgamation of two production organizations resulting in slimming

During cost cutting and downsizing, maintenance spending is often reduced with the justification that activities are only being "deferred." If key personnel such as maintenance coordinators and planners are lost at the same time, the organization may not have the institutional knowledge of which tasks are too critical to defer. Personnel who remain may be overworked as the organization tries to "do more with less." The longer the tasks are "deferred," the more likely it is to believe that we can operate like this for the long term. This can result in jobs which are poorly planned and executed.

Loss of key personnel from large projects can result in very similar situations with key knowledge being lost and remaining personnel too busy to give adequate thought and analysis to key decisions. If major projects *and* the existing facility are both understaffed, these problems can compound and increase the risk of a serious process safety event.

When it comes to accidents, organizational factors are conditions rather than causes. In risk assessment terms, they are enabling events. An example would be hazard warning signs that could have averted an incident are missed due to inadequate experience or lack of knowledge or skills. The impact of organizational change is often subtle and not immediately apparent. The effects can be degraded performance following increased workload or span of work (as shown in Table 5.1). The change may also have disrupted unrecorded or informal activities or communications that contributed to safety. This is more likely when baseline competencies for assigned tasks are not documented.

5.2 TASK ALLOCATION CHANGES

Some of the more common reasons for task allocation changes include:

- Changing job competency requirements
- Requiring individuals to take on new responsibilities demanding skills and competencies unconnected with those previously required
- Temporarily backfilling for someone due to leave of absence, temporary assignment, or extended vacation.
- Relinquishing an individual's responsibility for a task without that task being reallocated
- Temporarily not filling a position (e.g., hiring freeze)
- Organizational redesign

The first three causes listed involve placing new task demands on individuals who may not be fully equipped (skill-wise) to deal with the new requirements. The next two causes result in task assignments either temporarily or possibly permanently going unassigned. An organizational redesign can result in either or both of these consequences. In any of these cases, the change can result in either underperformance or nonperformance of a task. The risk of this event obviously depends on the criticality of the task to process safety. We will examine each of these types of task allocation changes in more detail in this chapter.

5.3 JOB COMPETENCY CHANGE

This type of change can occur when methodology or technology advancement results in new skill requirements to meet the task assignment. Some examples include:

- Adoption of risk-based inspection (RBI) to manage mechanical integrity
- Addition of layer of protection analysis (LOPA) in PHA to determine the safety integrity level (SIL) gap for safety instrumented systems (SIS)
- Upgrading process controls incorporating "smart " (self-diagnosing) components

Some of these changes have a broader reach than others. Incorporating LOPA into PHAs would primarily affect PHA facilitators, whereas significant changes to process control hardware could affect control engineers, instrument technicians, and operators. The following case study provides an example of a process control upgrade that changed the job competency required of board operators while their basic assigned tasks remained the same.

5.4 CASE STUDY: BAYER CROPSSCIENCE LLC – INSTITUTE, WEST VIRGINIA, USA (2008)

The incident, which occurred during startup in August of 2008 and killed two workers involved an explosion in a pesticide Residue Treater at the Bayer CropScience plant in Institute, West Virginia. The plant produced the pesticides methomyl and larvin from MIC. The design intent of the Residue Treater is to decompose waste methomyl in a heated solution of methyl isobutyl ketone (MIBK). Under normal operating conditions, dissolved methomyl and other chemical residues are fed into the preheated Residue Treater partially filled with solvent. The methomyl safely decomposes inside the treater to a concentration less than 0.5% wt. On the night of the incident, methomyl-containing solvent was pumped into the Residue Treater before the vessel was prefilled with clean solvent and heated to the specified operating temperature. A runaway decomposition reaction occurred that overwhelmed the emergency vent system and the resulting overpressure ruptured the vessel killing two nearby operators and injuring eight others.

The Methomyl unit was in the process of being restarted after a lengthy shutdown to replace the old carbon steel Residue Treater with a stainless steel vessel and to replace the Honeywell™ DCS with a system based on Siemens™ PCS7 technology. The switch over to the Siemens control configuration began with the Larvin unit in 2006, with startup of that unit in early 2007. Operators received considerable training (including practice on a process simulator) for the new system at that time and had become proficient with using the system on the Larvin unit. Management concluded that comprehensive formal training and practice using the new DCS on the methomyl process were unnecessary. They incorrectly assumed that the operator's experience with the larvin control system would be sufficient to run the new methomyl system.

5.4.1 Lessons Learned

The introduction of the Siemens control system significantly changed the interactions between the board operators and the DCS interface. While the Siemens control system contained features intended to minimize human error, the increased complexity of the new operating system challenged operators as they worked to familiarize themselves with the system and units of measurement for process variables that differed from those in the previously used Honeywell system. In particular, board operators told the CSB that the detailed displays in the DCS were difficult to navigate. Routine activities like starting a reaction or troubleshooting alarms would require operators to move between multiple screens to complete a task, which degraded operator awareness and response time.

The management incorrectly assumed that the methomyl board operators had become proficient from many operating hours using the DCS on the larvin unit. It was not recognized that even minor differences in operation challenge an unfamiliar operator unless the operator has had process-specific training for new equipment.

Methomyl board operators had minimal training for a few specific processes, but general training took place during the operators' shift as time allowed and was self-directed and self paced. Informal, on-the-job training intended to develop the necessary skills to run the system can lead to inappropriate or incorrect practices that become the norm in the absence of proper training tools and instruction.

The CSB report points to incomplete operator training for a new computer control system as a "critical omission" that might have allowed operators to recognize the problem much sooner. While not the incident root cause, the failure to recognize this change in job competency requirements was an enabling event that removed a layer of protection and allowed the incident to progress.

5.5 ASSIGNING NEW RESPONSIBILITIES

This assignment change typically occurs for one of the following reasons:

- Advancement of position (e.g., operator to shift supervisor)
- Acquiring more responsibility (combining engineering and maintenance functions)

Position advancement occurs at all levels in the organization. Some advancements are a result of attrition (e.g., retirement or departure) and some are by design (fast-tracking of managers). The new skill requirements can be both task-specific (becoming a maintenance coordinator) or more managerial such as leadership and people skills.

One of the major causes of acquiring more responsibility is the recent trend in industry to shrink the size of organizations to achieve the goal of "do more with less." With fewer people in the organization, key tasks need to be redistributed to those remaining. This situation may result in some individuals getting assignments by default (somebody has to do it), after the logical (based on skills and experience) reassignments have been made. The impact of downsizing is often an increase in workload, span of work, or both for individuals.

Another situation is the combining of previously independent departments such as engineering and maintenance under one manager. This usually results in either the engineering manager or the maintenance manager becoming the overall manager of the reorganized unit. In either case, the individual selected may not possess all the competencies needed to manage both disciplines.

5.6 TEMPORARY BACKFILLING

This is analogous to a temporary process change; the change is intended for a limited timeframe. As in the case of a process change, this change needs to be recognized and managed during the time it is in effect. As with a temporary process change, this type of temporary organizational change still requires a risk assessment, including an evaluation of the necessary skills and knowledge required to fill this position. The individual tasked with backfilling needs to be secure in what they are being asked to do. If the risk assessment determines that there is a gap in skill or knowledge, then training or other alternative measures should be taken to ensure that all safety issues are addressed during this temporary period.

5.7 VANISHING TASK ALLOCATIONS

As mentioned above, the following list of task allocation changes have different causes from those just discussed, but the outcome is the same.

- Relinquishing an individual's responsibility for a task without that task being reallocated
- Temporarily not filling a position (e.g., hiring freeze)
- Organizational redesign

Namely, some tasks that were previously assigned are no longer performed. This issue needs to be recognized and the process safety impact of this change evaluated. Developing an understanding of all of the safety-related tasks that a person's job entails is not always an easy task, especially if it isn't formally documented. It may also be unclear whether an identified task even has a safety implication. Identification of these tasks can be done through interviews with that person as well as with their peers and internal customers. Management must be alerted to all tasks which will not be reassigned, so each can be carefully assessed prior to elimination. In some cases, the loss of a previously performed activity may have significant safety implications as the following case study for relinquishing an individual's responsibility without reallocation of his tasks illustrates.

5.8 CASE STUDY: BP– WHITING, INDIANA, USA (1998–2006)

This example was uncovered by the U.S. BP Refineries Independent Safety Review Panel and involves a rupture disk deficiency at BP's Whiting refinery. Some time prior to 1993, rupture disks were installed under relief valves on the fractionators of two FCCUs. The rupture disks were installed either to protect the relief valves from contamination by process fluids that could impair their operation when required to operate or to eliminate emissions.

This type of installation has been shown to create a hazardous condition in the event that the rupture disk develops a pinhole leak that allows the pressure between the rupture disk and the relief valve to equilibrate with the process pressure. In order for the rupture disk to burst at its specified set pressure, the backpressure on the device needs to be low (typically atmospheric pressure). In the event of a pinhole leak, the rupture disk backpressure can rise until it reaches the equipment operating pressure, which effectively raises the burst pressure by a like amount. Depending on the set pressure of the rupture disk relative to the maximum allowable working pressure (MAWP) of the equipment, the equipment can be subjected to an overpressure that is unsafe. The typical layer of protection is to install a "tell tale" pressure gage between the rupture disk and the relief valve that is frequently monitored by operations. When pressure is noted, the rupture disk is replaced with a new one.

Between 1993 and 1998, the Whiting refinery executed numerous work orders to replace failed rupture disks. During this period, Whiting refinery personnel viewed the rupture disk replacement as an economic/reliability problem, not as a safety concern. In planning for a 1998 turnaround involving the two FCCUs, the refinery proposed to reengineer the system by replacing the relief valves and eliminating the rupture disks. However, the project was removed from the turnaround schedule, apparently due to budget and schedule pressures.

After BP's acquisition of Amoco in 1999, BP made significant staff reductions and reassignments at the Whiting refinery. As part of these actions, the refinery released the engineer responsible for reengineering the system and did not subsequently reassign or complete the rupture disk upgrade project. Between 2001 and 2005, the refinery completed only one work order to replace rupture disks. Each quarterly log dating back to at least April of 2004 indicated that rupture disks had failed (as evidenced by high pressure between the rupture disk and relief valve).

In 2005, after a review of open work orders by operations prompted an inquiry into the nonreplacement of failed rupture disks, the upgrade project was reinstituted and a new engineer was assigned. The prior project engineer was contacted and consulted regarding how to proceed, and he informed the new engineer about the hazard presented by failed rupture disks. The panel's technical consultant found the problem in 2006, before the refinery had completed the upgrade project.

5.8.1 Lessons Learned

While release of the first project engineer and not reallocating the upgrade project were not the main root causes (i.e., lack of understanding of the safety hazard by decision makers), it was an enabling event that arguably allowed the situation to go uncorrected for years. According to the panel's report, the engineer that was let go understood the safety implications of failed rupture disks under relief valves. Had the upgrade project been reassigned to someone else using an organizational change management system before the project engineer's departure, that urgency could have been explained to the individual who was reassigned the task.

5.9 IMPACTS/ASSOCIATED RISKS

Most organizational change management evaluations are done using brainstorming, checklists, What-If techniques, or a combination of both. Brainstorming is the least structured and most dependent on the knowledge and experience of facilitator and team. What-If is more often a blend of checklist and brainstorming. It is more structured and is typically started with some predetermined "What if" questions (checklist), which can then be augmented (brainstormed) by the review participants. The questions are addressed to determine the impact of the change. Table 5.2 gives some examples of What-If questions that are appropriate for evaluating possible impacts of task allocation changes. Additional checklists and What-If tools can be found in Appendix A.

The key objective of a risk assessment is to ensure that following the change the organization will have the resources (personnel, time, information, etc.), competence, and motivation to ensure safety without unrealistic expectations of people. Task mapping can be a very useful tool in these situations. Refer to Table A.16 in Appendix A for a checklist which may be helpful for task mapping.

TABLE 5.2 Task Allocation Change Impact What-Ifs

What If...	
1	The key EHS critical roles and responsibilities for the task have not been identified?
2	The individual assigned the task does not have the skills and experience to fulfill the key EHS critical roles and responsibilities?
3	The key EHS critical roles and responsibilities have not been communicated to the individual assigned the task?
4	The time allocated to the individual assigned the task is inadequate to perform the job safely?
5	Not all the key roles and responsibilities have been accounted for when the task was reallocated?
6	Informal safety activities or communication paths are disrupted due to reallocation of the task?
7	The policies, procedures, lists, or organizational charts have not been updated to reflect the task allocation change?
8	The emergency response function has not been reassigned when the position is eliminated?
9	There is an emergency during startup or when someone is on leave?
10	The risk of the impact on general EHS performance due to the task reallocation has not been evaluated?
11	The individual has not received proper training to carry out the EHS responsibilities?
12	A competency training matrix has not been reviewed and an appropriate training plan has not been created?

While the hazard HAZOP study methodology has established itself as a widely used technique for identifying and assessing process hazards, it has been applied to other nonprocess systems for evaluating safety, including the British railway system. By simply changing the parameters to which the action words (e.g., more, less, none, etc.) are applied, it can be adapted to many situations. Examples of modified parameters for organization change risk analysis are presented in Table 5.3. Another example can be found in Example B.4 within Appendix B.

The main action deviation words could be more or less—depending on the parameter.

TABLE 5.3 Examples of HAZOP Parameters for Organizational Change Review

Parameters for HAZOP	Parameters for Organizational Change Review
Flow	Work Flow
	Communication
	Networking
Pressure	Stress
	Work Load
Level	Capacity/Human Resources
	Span of Control
	Knowledge/Experience
	Authorization
N/A	Resource Proximity/Location
N/A	Logistic
Emergency Response	Emergency Response
Startup	Startup
Emergency Shutdown	Emergency Shutdown
Human Factors	Human Factors

5.10 CONCLUSIONS

Task allocation changes can occur as part of a larger organizational change or they can be considered a stand-alone change. Any time this type of change occurs it is essential to consider what happens to all of the safety-related tasks and who is responsible for each of them after the change. Task mapping is a good way to understand the complete impact of such a change. A risk assessment of the change may identify key actions which should be taken to ensure that all tasks are handled by competent personnel.

REFERENCES

Baker Panel, *A Case Study for Review of BP's Process Safety Management Systems*, 2007.

Castro, S.D.M. and Burnett, S., *Management of Organizational Change to Address HSE Risk - Effective Industry Practices*, Houston, 2010.

Center for Chemical Process Safety (CCPS), *Guidelines for Preventing Human Errors in Process Safety*, New York, 2004.

CSB Report, *Investigation Report, Pesticide Chemical Runaway Reaction Pressure Vessel Explosion*, August 28, 2008.

Health and Safety Executive Information Sheet, *Organisational Change and Major Accident Hazards*, CHIS7, Castleford, 2003.

Jacobsson, A., *Handling Downsizing in the Process Industries Experiences from the Swedish Process Industries*, IChemE 12th International Symposium on Loss Prevention, Edinburgh, 2007.

Reason, J., *Managing the Risk of Organizational Accidents*, Ashgate Publishing Company, London, 2004.

6

ORGANIZATIONAL HIERARCHY CHANGES

Reorganization or organizational redesign can come from a number of directions, but the starting point is nearly always the need or opportunity to improve a company's financial performance. Indeed the company's financial survival may depend on how effectively and quickly this change can be executed. While it is sometimes viewed as rumblings at the top which have no effect on the shop floor, this is seldom the case. By its very nature and purpose it is being done to have just such effects. For example, when a company delayers its hierarchy, it brings those "rumblings at the top" closer to the shop floor. Layers are no longer there to dampen or filter the change in how the day-to-day shop floor work proceeds.

How the shop floor work proceeds (i.e., what happens in the control room, laboratory, engineering office, or maintenance shop) can either prevent or cause a process safety incident to occur. Hierarchy changes can either help or hurt process safety. Which way the change affects process safety depends on how it is led, managed, and designed. Process safety will not take care of itself, cannot wait till later, and should not be a secondary concern. Indeed during the time of "confusion" resulting from a hierarchy change, manufacturing operations are often at their greatest vulnerability to having a process safety incident. For this reason, process safety, along with personnel safety, must be leading concerns instead of lagging incident reports.

For those primarily concerned with the financial costs and opportunities from the planned change, it is good to note the severe financial consequences of process safety incidents, such as the Deepwater Horizon incident. According to a CCPS study, in the

U.S., major industrial accidents cost an average of $80 million each and business interruption costs can amount to four times the cost of the property damage from an incident.

Given these warnings but also recognizing that change must and will occur, this chapter will examine some of the more common types of hierarchy changes that occur today and the process safety concerns and issues triggered by such changes.

Because there is generally greater vulnerability during organizational transition, special consideration should be given to process safety during and immediately after the transition occurs. Keep in mind that not only is a business being changed but people's lives are as well. Some of the more common personnel and process issues which should be dealt with during the transition will be identified and discussed in this chapter.

Hierarchy changes can also offer the opportunity to improve process safety. Some of these opportunities will be identified and discussed in this chapter as well.

6.1 CENTRALIZATION OR DECENTRALIZATION OF JOB FUNCTIONS

For various business reasons companies may centralize or decentralize various functions. Examples may include centralization of R&D to facilitate researchers working together or decentralization of project engineering to allow localized control of capital spending. The size of manufacturing locations and financial considerations may necessitate the sharing of certain overhead functions and therefore centralization. Generally, direct operations and maintenance functions are less likely to be centralized. Process safety can be affected by these decisions as the following example illustrates.

6.2 CASE STUDY: ESSO – LONGFORD, VICTORIA, AUSTRALIA (1998)

In 1998 at an Esso gas plant in Longford, Victoria, Australia, a heat exchanger fractured, releasing hydrocarbon vapors and liquids. Explosions and a fire followed, killing two employees and injuring eight others. T. Kletz, in his book *Still Going Wrong* offers the following observations concerning this incident:

> ...all the engineers, except for the plant manager, the senior man on site, were moved to Melbourne. The engineers were responsible for design and optimization projects, and for monitoring rather than operations. They did, of course, visit Longford from time to time and were available when required, but someone had to recognize the need to involve them.

6.2.1 Lessons Learned

Kletz goes on to say:

> The physical isolation of engineers from the plant deprived operations personnel of engineering expertise and knowledge, which previously they gained through interaction and involvement with engineers on site. Moreover, the engineers themselves no longer gained an intimate knowledge of plant activities. The ability to telephone engineers if necessary, or to speak with them during site visits, did not provide the same opportunities for informal exchanges between the two groups...

Here is a good example of centralization being one of the factors in a process incident. Engineering was centralized and three significant issues emerged:

- The operations department was deprived of day-to-day technical monitoring and support.
- Engineers lost intimate knowledge of operations.

- Informal communications between engineering and operations broke down.

Even if these people had not been relocated, these same issues can occur. When people become part of a centralized function, they can lose their sense of responsibility and cease to feel they are an intimate part of the production team. When they are no longer considered a direct team player, but instead a resource to be called upon, a number of communication channels can disappear. The adverse effects of this centralization can be reduced by assigning certain people to be key liaisons for certain production facilities or departments and encouraging them to remain “in the loop” with the day-to-day operations.

These are technical and operational issues that a company should address prior to the centralization. In this case, not effectively doing so had a disastrous impact on process safety.

There are some good reasons to centralize which in some cases may even enhance process safety. With the ever-growing complexity of technology, it may not be possible or reasonable to have experts in every field at every manufacturing location. The experts may not even have enough work at a given location to make it a full-time job or to become proficient at their specialty. There can also be the need to develop personnel through varied experience at multiple locations. Centralization can help facilitate technical exchange between manufacturing locations and combat the “not invented here mentality.” This can be particularly important to process safety. While each plant should have a robust process safety program that works at that location, the corporation should have plants that adopt the “best practices” from one another. Central groups who serve all plants are often the catalyst that makes such adoption possible.

Obviously there are pros and cons impacting process safety through changing an organization in either direction. If done correctly, either can facilitate process safety. Prior to making the centralization or decentralization decision, its impact on process safety needs to be determined. The resulting organizational change can then be designed to move process safety in the right direction.

6.3 REORGANIZATIONS AND DELAYERING THE HIERARCHY

Reorganization and delayering is common in today's corporate world because if done correctly it not only lowers cost but potentially offers the following advantages.

- Better communications result from a reduction in hierarchy. Fewer levels through which upper management must communicate allows it to deliver its message clearer and with less "editing" as it goes down the organization. By the same reasoning, messages come up the organization to upper management with a similar reduction in editing.
- The company becomes more nimble, able to adapt and respond to changing economic and market conditions more quickly.
- Improved relations between upper management and employees may be realized. There is greater opportunity for personnel to know and communicate with upper management: who they are, what they want, and what their goals are. By the same token, upper management gets to better know employees: their needs, career aspirations, and concerns. In other words, the ivory tower has fewer floors.

When planning a reorganization, a company should ensure that no gaps are left in management and application of process safety elements. There are many elements to a strong process safety management program. They are all necessary, and they all need at least one person responsible for ensuring they are being done effectively. These people are often referred to as subject matter experts (SMEs). Each SME should be qualified and clearly charged with the task and responsibility. Furthermore, everyone else should know that the SME is the "go to" person for the particular process safety element. Now this may seem so obvious as to not require mentioning, but unfortunately it is often overlooked. It may be something as basic as determining who is qualified to write a hot work permit or who can evaluate a pressure relief valve. It might also be which new manager has overall process safety responsibility.

The plant management gives the assignments for process safety responsibility to all affected employees, contractors, and support organizations so that they know who has which process safety responsibility. There are lots of ways to do this but the more visible the better.

One company addressed this issue with the following actions. The production unit posted on a white board at the entry to the unit the process safety job responsibility with the assigned SME's name. Since some of the names were different with each shift change (for example, permit writing) some of the line-up changed each time a new shift started. Updating the board may sound like a cumbersome task, but questioning the staff, supervisors, engineers, technician, craftsmen, and operators to find who could write a permit takes even more time. More importantly, it gives a clear picture to everyone who manages process safety, how they manage the critical tasks, and most of all, how important it is.

Using the shift change as an analogy there is a shift change occurring when the company reorganizes. Process safety duties are changing and some of these duties may be temporary. Publishing even the temporary assignments lets the organization know that management has not forgotten process safety even if it too is in transition.

Immediately after the new organization is announced is a great time for the new leaders to get out in the plant and affirm their commitment to safety, of which process safety is a key part. The first thing addressed by management will be what most people subsequently believe is the first priority of management. If the first issues addressed are personal safety and process safety commitment, the new management team cements these values into the new organization. Although these first impressions set the tone, this commitment to safety needs to be reiterated frequently. New management also needs to be careful not to overload the organization with too many priorities or they will all get lost in the tide.

Good communications involve listening as well as speaking. Asking the right question in the right way can be much more effective than any written survey. The manager has a chance to read the body language, the voice inflections, and other nonverbal communications of the employee speaking. He has a chance to be warned about what may have been overlooked, misinterpreted, or misunderstood.

This is a real opportunity to get the new organization started in the right way. It is also a chance to start the discussion of a topic upon which both management and labor can agree. Good process safety is a win–win for everyone. It is in everyone's best interest to make it work.

6.4 IMPACTS/ASSOCIATED RISKS

The reduction of management levels causes reductions of personnel, often including personnel that have been good performers. If, for example, a whole layer is removed four levels up from the shop floor, these are probably people who over a period of time have been promoted because they are good performers. They may now be terminated from the company or bump someone else who is also a good performer. These personnel reductions create personal stress for each individual in the organization. This can cause anxiety, anger, depression, and lack of focus, which can adversely affect process safety. This issue is dealt with in Chapters 3 and 4 and therefore will not be repeated here.

Levels of management eliminated are not always elevated positions. For example, a company may decide to eliminate shift supervisors in favor of self-directed work groups. In the old organization, the shift supervisors would normally train the new operators. So in the new organization, training should be assigned to someone else. This seems obvious when so stated, but in practice important details such as this may be overlooked.

The term "corporate memory loss" refers to the loss of knowledge during reorganization. Often downsizing is done using attractive retirement packages, thereby resulting in a disproportionate loss of experienced personnel. Such persons may include, among others, shift supervisors, subject matter experts, process engineers, design engineers, and maintenance personnel. What these people know should be passed on before they leave. Nowhere is this more important than process safety. Ideally this is an ongoing process since often a person's interest in passing along this knowledge diminishes quickly once they decide to leave or they are informed that they are leaving. The ability to detect, troubleshoot, and correct upset conditions is essential to preventing unintended releases and other process incidents. Lack of this ability has been a major factor in some of the more serious incidents that have occurred.

Task mapping is a good tool for ensuring that the new organization will have "all the bases covered." Checklists can also be helpful. Refer to Appendix A for additional information on these tools for risk assessments.

When the organizational structure is being changed or it has recently been changed, and personnel don't fully understand it, there is an unsettled period within the organization. There are new people performing new jobs with which they are not yet comfortable. Functional questions arise such as, "Who writes the hot work permits now that the shift supervisor is gone?" Some departments or the entire plant may be understaffed if more people than planned accepted the early retirement package. Employee morale is low and people are worried about losing their job. All these are negative circumstances which may be in play during reorganization. During a production turnaround, process safety issues abound. Process changes are made, there is a high level of maintenance work occurring, emergency relief systems can be undergoing evaluation and repair, and nonroutine work requiring permits and special procedures is being done. All of this requires a robust process safety program. In the active stage of organizational restructuring and immediately after the change is completed is not the time to enter into a complex activity such as a turnaround.

6.5 CHANGES TO SPAN OF CONTROL

A topic closely related to delayering is a change to the span of control. Sometimes this change occurs as a result of a significant reorganization, but it can also happen with simpler organizational changes. In this type of change, the scope of someone's job responsibilities is changed. If the span of control is increased the types of responsibilities are the same, but the person may be expected to provide oversight, guidance, and expertise to more people or departments within the organization. If not handled correctly, this type of change can result in someone being "spread too thin." If the span of control is decreased, then there may be tasks or areas which are no longer adequately assigned. (Refer to Chapter 5 for discussions of task allocation changes.) During any change to a span of control, it is essential that the responsibilities are clearly communicated throughout the organization so that everyone understands who is in charge of which tasks and areas.

6.6 IMPACTS/ASSOCIATED RISKS

It is important to understand the demands of a job on a person's time before increasing the span of control. If the change to the scope of a job means that the person will be less available to the people who rely on them, then it may encourage people to find alternative options. On the positive side, this can sometimes empower people within their own jobs and require less assistance. However, it can also mean that people conduct their business with a lack of information or rely on incorrect sources which can lead to catastrophic errors.

Checklists and "What-If" brainstorming techniques could be useful in assessing a change in the span of control. Refer to Appendix A for additional information about these risk assessment tools.

6.7 LINEAR VS. MATRIX ORGANIZATION

In the business world, there are many different types of organizational structures. Most structures have changed in an evolutionary process to meet the changing needs and/or opportunities that a company faces. Sometimes this trends toward more complexity, moving in the direction of a matrix structure and away from a branching structure resembling the ancestral family tree. To a large extent the move toward a matrix structure is recognition that no one individual can be expert or even minimally competent enough with the myriad of functions that a modern company must do. These many functions have increased the need for specialists and teams. Indeed many companies are even eliminating generalist jobs.

Effective process safety is a good example of a business function that generally needs specialists. For instance, not everyone can be or needs to be an expert in pressure relief and flare system design—but someone needs to be an expert or SME in the field.

Additionally process safety requires teams. Teams are often brought together to perform PHAs or hazard identifications, PSSRs, or to review MOC plans. Teams are sometimes used to write and/or review operating procedures. Teams facilitate bringing together the SMEs assigned to a task. Here SMEs can interact to be sure all aspects of the task are addressed. Further peer review teams may be used to ensure that nothing important is overlooked, misinterpreted, or miscalculated. These team activities are consistent with a matrix structure.

Another function of a team may be acting as a bridge between developing or changing a PSM element and getting it implemented. Operations and maintenance personnel often serve as the best editors for procedures developed in a team. Further, teams can be used to gain the buy-in from those who are affected by, or have to implement, the change. Employee involvement is a key element of process safety. Nothing addresses this as well as employee involvement in developing, editing, reviewing, and implementing process safety procedures.

Effective teams require a common goal and mutual respect. Process safety focused teams should meet these criteria. Everyone in the plant benefits from good process safety management; it is a common goal. Respect comes when team members realize the importance of each member's contribution. The contribution generates the respect it deserves, and since most people want to be respected, they excel in such an environment.

6.8 CASE STUDY: BP—TEXAS CITY, TEXAS, USA (2005)

The Report of the BP U.S. Refineries Independent Safety Review Panel, frequently called "the Baker Panel Report," made the following observation:

> Accountability for process safety performance at BP's U.S. refineries is with the line that runs from (a) the Group Chief Executive to (b) the Chief Executive, Refining and Marketing to (c) the Group Vice-President, Refining, whom BP considers the most senior manager of the refining business to (d) the Refining Vice President-North America to (e) the individual refinery plant managers to (f) operating personnel at the refineries. Outside of line management BP has a number of functions and positions with some connection to process safety in the refining business. Currently, process safety functions and expertise reside with a number of different parts of the BP corporate organization, including
>
> - the BP Group Technology function
> - the newly formed Safety and Operations function
> - the Refining Technology Group
> - the Group Vice-President – HSSE (EHS) and Technology for Refining and Marketing
> - the HRO manager in the refining organization
> - the new position of Vice President of OMS/PSM program Implementation
> - the Chief Executive Officer of BP America
>
> The panel believes that this organizational framework produces a number of weak process safety voices—weak in the sense of apparently lacking the ability to influence in a meaningful way strategic decision-making with respect to U.S. refining operations.

6.8.1 Lessons Learned

The report presents concerns from BP's organizational structure that points to the weakness of both hierarchal and matrix structures. At BP, process safety accountability ran in a line from the Group Chief Executive to the Refinery Plant Manager. The problem is that this is only one of many competing accountabilities. Other accountabilities included, "all aspects of performance in the refineries, including financial, budgetary, environmental, and otherwise." The report goes on to say that process safety received low priority due to competing objectives.

The report also presents concerns directed at matrix structures. While there are seven key individuals and departments with process safety interest, these lacked the "ability to influence" process safety. Many teams and departments don't solve a problem if they do not have the authority and resources to do so.

Since both structures have problems, what should a company do? The Baker panel had a simple answer:

> The Panel believes that process safety in BP's refining operations would benefit from BP's designation of a high-ranking process safety leader who will participate consistently in important refining decision-making that affects process safety performance.

This is certainly a good start. With the right mandate such a position goes a long way toward solving the issues identified. If this person has the knowledge necessary to implement process safety, or can draw that knowledge from his/her section of the matrix, and the ability to influence the main hierarchy with regard to strategic decision making for process safety then the company can avoid the pitfalls of these types of organizational hierarchies.

Regardless of the type of organizational structure that is utilized, it is essential that each person tasked with process safety responsibilities have the authority, resources, and priority to act. Process safety can function very well in a company's organizational structure, whether it is a matrix or linear arrangement, as long as it is considered an important business function and given the required visibility, commitment, funding, and personnel. Like any business function, process safety needs its place in the management structure.

6.9 IMPACTS/ASSOCIATED RISKS

Both linear and matrix structures can be used to effectively manage process safety but there are pitfalls with each which must be avoided. Here are some of the issues that should be considered when contemplating a change to your organizational structure.

"If it's not broken, don't fix it"—this axiom is often good advice but not universally true. Any significant change must weigh the benefits against the detriments. It must also consider the disruption as both short term and long term. There can however, be opportunities to take process safety to a whole new level that can only be reached by significant change in the organization structure.

Centralize at a manufacturing site instead of headquarters: Many companies do this. It eliminates a lot of travel and keeps the centralized function close to one operation. Oftentimes the host site becomes an incubator for new ideas. Research departments learned this a long time ago. The same advantage may be gained for process safety.

To meet a time constrained need, temporarily centralize: This is often done for a capital project. Other work of a project nature may justify this approach.

Use technology to improve communications: The ability to share real time data and conduct effective meetings with people in remote locations has changed the paradigms of how business is conducted. Physical location has become less important and much less of a hindrance to effective businesses.

Foster, and even insist on, best practices exchanges: "Not invented here" has no place in an effective organization. No place is this more true than it is with process safety.

When evaluating the risks associated with altering the type of organization, a “What-If” brainstorming approach may be the best option. Refer to Appendix A for additional information regarding risk assessment tools.

6.10 ACQUISITIONS, MERGERS, DIVESTITURES, AND JOINT VENTURES

This is a definite and special case of reorganization. It is definite, because it will nearly always result in organization structure changes. It is special, because the change is likely to be significant. The resulting company will have a different organization than the original company or companies did before the change. The new management team will be different. The business plan can change. Hopefully such things as revenue and profitability will be improved.

The change is likely acute in its timeframe. The new company must function immediately after being formed. There should be at least a temporary structure which makes this possible. In most cases, manufacturing operations continue to run during the transition and so should process safety. There is no holiday for safe operations.

6.11 CASE STUDY: ANONYMOUS, USA (1998)

An electrical fault occurred in the motor control center while operators were attempting to restart the chiller compressor. The electrical fault caused the main breaker for the motor control center to trip. The tripped breaker shut down several critical pieces of equipment for the production line. Operators investigated but were unable to start the critical equipment they needed and subsequently followed the emergency shutdown procedure to safely shut down the line. The fault was caused by a broken wire which was a result of inadequate maintenance due to a failure of the work order system.

During the transition period after being purchased by another company, the site was relying on the previous owner for preventive maintenance services. A thermography report issued eight months prior to the incident indicated the failed breaker as a potential problem. However, no work order was issued to address the breaker because the thermography technician did not have access to the site's work order system.

6.11.1 Lessons Learned

During a change in ownership, companies need to ensure that all aspects of process safety are functional and that everyone understands their roles, responsibilities, and the resources available to get their jobs done completely and correctly. If there are gaps in systems due to the change in ownership, these need to be identified, and adequate temporary systems put in place until the organizational change is complete.

6.12 ASSOCIATED RISKS

The change is likely made with less data than anyone would like to have. Due diligence is usually time- and data-constrained, but it is critical that process safety is a part of the due diligence process. Indeed, it is one of the very important parts of due diligence. Poor process safety can be a significant liability affecting the cost and success of the acquisition or merger. According to the CCPS *Guidelines for Process Safety Acquisition Evaluation and Post Merger Integration*, "Poor post-deal integration was identified as the major factor or primary reason that, in as high as 43 percent of mergers, the merged company or operations did not outperform competitors."

When blending roles like this, task mapping may be a good option to ensure that there are no critical gaps in the new organization. The task mapping can also be a good way to communicate new responsibilities within the revised organization.

6.13 CASE STUDY: UNION CARBIDE—BHOPAL, INDIA (1984)

The liabilities can have lasting long-term consequences as shown by the following case study from the Bhopal, India incident.

In M. J. Peterson's "Bhopal Plant Disaster – Situation Summary" he states:

> Locally and globally, blame for the accident was quickly assigned to Union Carbide. Consistent with widespread beliefs that multinationals control their subsidiaries' operations very closely, it rather than UCIL, was deemed ultimately responsible for the condition of that plant and the level and training of staff.

The Bhopal disaster occurred in 1984. In 1989, an Indian civil court ordered Union Carbide to pay $470 million. In the early 1990s, Union Carbide divested itself of its stock in the Indian joint venture, UCIL, which had owned Bhopal. In 2001, Dow acquired Union Carbide. In December 2010, Dow was hit with a $1.1 billion claim from India's attorney general for victim's compensation and environmental cleanup costs. The claim will be decided by India's Supreme Court.

6.13.1 Lessons Learned

Twenty-six years after the Bhopal incident, its impact continues to be felt by Dow. Interestingly, Bhopal began with a joint venture between Union Carbide and Indian investors. This, however, did not ameliorate Union Carbide's liability. For a company to assume these liabilities it should have a process safety organization in the subsidiary through which it can implement, control, direct, and improve process safety.

A more detailed treatment of mergers and acquisitions can be found in the CCPS publication *Guideline on Process Safety Consideration during the Merger and Acquisition Process*. Quoting the guideline:

> ...a number of the member companies of the Center for Chemical Process Safety investigated the costs they were incurring to rectify process safety issues found after they had closed on an acquisition. Arising out of their investigation they found it was common that another ten to thirty percent of the initial purchase price was being expended rectifying such issues.

Clearly, process safety performance is important to both risk avoidance and financial performance. Make process safety a key part of due diligence. The due-diligence team needs a process safety representative to be sure the right information is requested and once the information is received that it is properly evaluated. Oftentimes the evaluation centers around performance and liabilities; incidents, inspections, lawsuits, fines, etc. These are important lagging indicators. The leading organizational indicators are also important. Some of these are:

- Who is responsible for each element of process safety management?
- Is there evidence that responsibility for all process safety elements has been assigned to trained personnel or a SME?
- Is the assigned person in a line or a staff role?
- How are process safety decisions made and implemented?

- Do the responsible personnel have the authority and funding to carry out process safety?
- Will the responsible parties come with the acquisition or remain with the divesting company or even leave both companies?

These are key organizational questions that may not be fully answered in the due diligence process. It is to the best interest of both parties in the acquisition, merger, or joint venture that there be continuity for process safety and this requires transparency of the existing management process.

Decisions regarding how to integrate process safety within the new organizational structure should be made, taking into consideration the strengths and weaknesses of the original companies involved. In some cases, the acquiring company may have little or no experience managing PSM-covered facilities. Examples may include:

- A private equity firm making its first refinery or manufacturing facility purchase
- A marketing or nonmanufacturing company purchasing an upstream chemical or polymer facility to secure feed stock or raw material
- A subsidiary owned by a chemical manufacturing company in partnership with banking or private equity firms

For these facilities there should be good understanding of process safety. Senior management needs awareness and understanding of process safety. It may need to call in an outside consultant specializing in process safety to conduct a PSM audit and/or give management advice about process safety. Senior management may also decide to make a process safety manager part of its team or have a SME report to advise them. The legal structure of the partnership should empower the most-qualified and experienced partner the authority and latitude to create and manage process safety programs.

6.14 CHANGING SERVICE PROVIDERS

Sometimes companies need to outsource various aspects of process safety management, such as maintenance. The initial decision to contract out such activities as well as the choice to change service providers should be considered an OCM. More information regarding the initial decision to outsource process safety functions is covered in Chapter 7.

Contracting out a process safety function can introduce many complications in terms of reporting structure, authorizations, and responsibilities. Even if you have these issues ironed out with one contractor, when you change service providers the same issues should be revisited with the new contactor. It is often necessary to have very specific written guidelines for how this process safety function will be managed, which may be different than if the function were handled by in-house staff.

6.15 IMPACTS/ASSOCIATED RISKS

As noted in the case study in Section 6.11 regarding the failed work order system, there can be issues with access to key systems necessary to perform the complete job, including the associated paperwork. Another possible concern is ensuring that the contract employees are properly trained and then determining who is responsible for maintaining and auditing this training documentation.

Checklists and task mapping would be appropriate risk assessment tools to assist in evaluating these types of changes. Refer to Appendix A for additional information about these tools.

6.16 CONCLUSION

The hierarchy of a company is created to best carry out the business purpose of that company. The hierarchy changes because something about the company has changed or needs to change. Process safety can be made to flourish in most structures as long as it is well thought out. Keys to changing the hierarchy and maintaining and improving process safety are:

- Recognition that when there are significant organizational changes, personnel can be distracted from focusing on process safety unless the transition plan takes this into account
- Good communications before, during, and after the change are critical
- Personnel should be trained in process safety
- There should be a structure for carrying out process safety which exhibits clear assignment of responsibility
- Senior management should understand, lead, and support process safety
- There should be financial and personnel resources available to carry out the process safety strategy
- The company should have dedication to process safety excellence and improvement

REFERENCES

Baker Panel, *A Case Study for Review of BP's Process Safety Management Systems*, 2007.

Center for Chemical Process Safety (CCPS), *The Business Case for Process Safety*, 2nd Edition, New York, 2006.

Center for Chemical Process Safety (CCPS), *Guidelines for Acquisition Evaluation and Post Merger Integration*, New York, 2010.

Kletz, T., *Still Going Wrong! Case Histories of Process Plant Disasters and How They Could Have Been Avoided*, Gulf Professional Publishing, Houston, 2003.

Petersen, M.J., *Bhopal Plant Disaster - Situation Summary (Draft)*, International Dimension of Ethics Education in Science and Engineering Case Study, March 2009.

Wall Street Journal, *Dow Chemical Is Hit with Demand for Bhopal Payouts*, New York, December 3, 2010.

7

ORGANIZATIONAL POLICY CHANGES

While the preceding chapters have dealt with issues related to changes in personnel or assignments within an organization, this chapter considers policy changes. Although these changes may not be associated with immediate organizational changes, they can serve as enabling events which could adversely impact process safety within the organization. Because these types of changes do not involve actual changes to the people or tasks within the organization, they can easily be overlooked by an OCM procedure, but this chapter will highlight why these changes should be included within the scope of an OCM process.

7.1 CASE STUDY: DUPONT—DELAWARE, USA (1818)

During the 19th century, DuPont sought to ensure workplace safety through both formal rules and personal stewardship. Safety rules had been put into writing and circulated widely by 1811. After an 1818 explosion, the members of the DuPont family, all of whom were away during the incident, agreed that one partner should always remain in the yards and that lower-level managers reside, as they did, on plant grounds. A ban on drinking alcohol, which had been implicated in the disaster, was also instituted. Technological improvements were also pursued with an eye toward improved safety. When Lammot DuPont moved to involve the company in dynamite production during the 1880s he realized production risks would be greatly increased. He had hoped that mechanization at the Repauno plant would enable workers to avoid the most dangerous types of work, but he died before accomplishing that goal. (Lammot DuPont died in an explosion at the Repauno plant.)

7.1.1 Lessons Learned

Here is an early example of organizational policy change addressing process safety. The DuPont family home and the gunpowder plant were on the same Brandywine River site. At that time, gunpowder plants frequently exploded and were generally located in remote locations so that only factory workers were impacted. Not only having their home so situated, but also establishing a policy "that one partner should always remain in the yards," spoke to the DuPont's commitment to improving process safety.

In the time in which all this took place (early 1800s); establishing safety rules was probably just as important as a policy change today. Safety rules have evolved into the policies and laws that govern process safety.

Like other areas of organizational change, corporate policy decisions should be examined for their effect on the company's process safety efforts. As in the above case study, companies can make an overt policy change to improve process safety. Although no reasonable company would purposely make a policy change that hurts its process safety performance, it is possible that a policy could have unintended negative consequences. This chapter will examine both the positive and negative effects on process safety caused by some policy changes which can affect process safety.

7.2 CHANGES TO MISSION AND VISION STATEMENTS

A company's mission statement generally describes in one to three sentences what the company does, its values, and what it wants to accomplish. The authors typically are the company executives or board of directors who set the goals and direction of the company. The target audience includes potential and existing investors, employees, government agencies, advocacy organizations, suppliers, customers, communities in which it operates, and the general public. It is no small task to communicate this concisely. Since everything that a complex company does cannot be covered in such a succinct statement, most companies attach further explanation in the form of corporate strategy, vision, values, or objectives.

Where does process safety show up in these mission statements and explanations? The current mission statements of nine major chemical and oil companies were examined [DuPont, British Petroleum (BP), Dow, Exxon, BASF, Chevron, Koch, Valero, and Bayer]. None of them mentioned process safety specifically. In the further explanation documents, some companies mentioned safety and environmental goals but there was no specific mention of process safety. This may be more understandable than it first appears. The audience is very broad, essentially everyone who is interested in the given company. Further, most people would not understand what process safety is, though some communities in which these companies operate understand very well what it is. It may also be that companies believe that operating without a process release is a given expectation in line with such basic expectations as accurate financial statements or obeying applicable regulations and laws.

These mission statements and associated documents set the tone for the company. Even if these documents don't specifically mention process safety, these high-level managers should find a way to convey the importance of safety for the company. The first recommendation from the Baker Panel report on the BP U.S. Refineries highlighted this need:

> "RECOMMENDATION #1- PROCESS SAFETY LEADERSHIP
>
> The Board of Directors of BP p.l.c., BP's executive management (including its Group Chief Executive), and other members of BP's corporate management must provide effective leadership on and establish appropriate goals for

process safety. Those individuals must demonstrate their commitment to process safety by articulating a clear message on the importance of process safety and matching this message both with the policies they adopt and actions they take."

The mission statement may not be the best venue for a company to communicate to all its stakeholders its commitment to process safety, but there needs to be a venue with comparable visibility which does communicate this commitment. The American Chemical Council's Responsible Care program was designed to effectively communicate a company's commitment concerning safety, the environment, and product stewardship. Prominent in the Responsible Care program is process safety. Companies which embraced and publicized their proof of commitment could rightly show that they have process safety at the forefront of how they do business. Tools such as community advisory panels, employee information meetings, and town hall meetings all are proven, effective methods that a company can use to tell its story.

Whenever a company decides to change its mission statement or other associated corporate policies, they should be reviewed for potential adverse impacts on process safety. As stated above, process safety may or may not be specifically mentioned in these documents, so it is essential to "read between the lines" to see if the changes might inadvertently reduce the perceived importance of process safety within the organization.

7.3 NEW AND REVISED CORPORATE PROCESS SAFETY RELATED POLICIES/PROCEDURES

Many companies have high-level policies and procedures which describe how they will address the many facets of process safety management. These are typically driven from a centralized corporate function and are applicable to all operating facilities within the company. On the surface it may seem like these are just typical policy changes which do not require an OCM. However, the specifics of the policy directives that are being made should be reviewed for potential impacts on the downstream organization.

In some cases these policy changes could end up requiring changes to someone's span of control or adding additional tasks to their job function for which they are not currently equipped to handle. As mentioned in earlier chapters, these types of changes fit within the framework of OCM if they have the potential to impact process safety.

If these new or revised policies require additional activities or documentation beyond what is currently expected, there may be significant resource burdens imposed on plant sites to comply with the new policy. These new resource demands may only be temporary, but they are real and need to be considered. In some cases, it may be necessary to provide for a temporary grace period during which it is acknowledged that the plant sites may not be in complete compliance with the new policy.

7.4 MAJOR CHANGES TO POLICY OR BUDGETS FOR MAINTENANCE OR OPERATIONS

These changes can take a variety of forms. To list a few:

- In the face of declining sales volume reducing headcount to maintain the same employee cost per unit of production (or productivity)
- Using a just in time (JIT) method to reduce maintenance spare parts inventory
- Installing self-directed work teams to replace hierarchal structure

- Changing the way costs and profitability are used to evaluate manufacturing
- Changing from preventive to run to failure maintenance
- Changing a company's business focus, resulting in deemphasis of the manufacturing operations
- Across-the-board budget cuts, especially if these cuts are compounded on top of cuts made over preceding years
- Reduced travel budget resulting in less personal contact
- Reduced training budget, training via web conference or computer-based training instead of classroom based

Any of these policy changes can be made and have no negative effect on process safety if it is clear that process safety is not an optional program which can be cut, delayed, or unfunded. It is a cost of doing business, but in both human and financial terms inadequate process safety is a much greater cost.

7.5 IMPACTS/ASSOCIATED RISKS

The risks associated with changes to policies or blanket changes to budgets is that the changes eliminate too many resources (people and/or money) resulting in an inability to properly carry out the necessary functions of a good process safety program. Although managers like to convey the message that everyone will just have to "do more with less" or "work smarter, not harder," there is a minimum resource level below which a facility cannot safely operate. As we discussed in some of the earlier chapters, it is necessary to determine the minimum safe staffing levels for each control room; a similar exercise is necessary to determine the minimum safe resource level for maintaining a good process safety program. This may be a little more difficult to do, since the tasks may be spread out among many different departments. However, unless a company has a clear understanding of what this minimum resource level is, they will not be able to adequately argue against budget cuts that go too deep.

Depending on the nature of the policy change or budget cut, there may be suitable checklists which can be utilized to evaluate this type of change. Otherwise a brainstorming or "What-If" technique may be appropriate for this type of risk assessment. Refer to Appendix A for information about risk assessment tools.

Before any major changes are made to policies or budgets, the potential impacts on process safety should be assessed. Thinking of this after a major incident caused by a policy change is too late. Such an audit should attempt to identify and deal with both intended and unintended consequences.

One cause for unintended consequences is often misinterpretation of the policy change. How the policy change is foreseen to impact, or not impact, process safety needs to be clearly understood and communicated. This communication involves both speaking and listening. Management leading the policy change should clearly communicate its intention prior to and during implementation. Management should also listen to feedback during and after implementation to be certain that desired results are being obtained. When a serious process incident is attributable at least in part to a policy change, this feedback part of the communication process is often found to be missing or inadequate. Too often one reads in an incident investigation report that employees knew the policy change was not working, but the decision makers did not know. Sorting out a natural resistance to change from substantive shortcomings in the changed organization can be difficult.

The sheer volume of policy changes can also be a problem. Quoting from the Baker Panel report:

> In addition, BP's corporate management mandated numerous initiatives that applied to U.S. refineries and that, while well intentioned, have overloaded personnel at BP's U.S. refineries. This initiative overload may have undermined process safety performance at the U.S. refineries.

It is the job of management to prioritize and manage all work, especially new initiatives, so that "overload" does not occur. It is the job of those developing initiatives and regulations to recognize an ever-increasing burden does not improve financial performance or safety. Good process safety is not a bureaucratic exercise. It is a carefully designed, lean, and clearly communicated process which can be effectively implemented.

7.6 IN/OUTSOURCING OF KEY DEPARTMENTAL FUNCTIONS (e.g., ENGINEERING DESIGN OR MAINTENANCE)

The OSHA PSM contractor element requires that employers obtain and evaluate information regarding the contract employer's safety performance and programs. Many employers use this as an opportunity to screen contractors and only hire and use contractors with safety performance records that meet their minimum criteria.

Anytime the sources of key services are changed, process safety can be affected. In the long term, it can and should be improved. In the short term, good management of change needs to occur. Engineering and maintenance are two key functions where this type of change often occurs and sometimes adversely affects process safety. Some of the risks or deficiencies which can occur are:

- Loss of knowledge gained from experience
- Loss of shared process safety values between the operating company and contract service provider
- Unclear reporting relationships concerning process safety responsibilities and authority
- Loss of direct authority over employees (i.e., a problem should be addressed by the plant owner through the contractor's chain of command)
- Loss of skills or expertise needed for specific functions

Experiential knowledge can be lost whether a function is in- or outsourced. Whether it is the operating company or the contractor, whoever is currently doing a job has generally obtained a useful knowledge base. A complete change-out of the personnel by altering whether the service is done by in-house personnel or a contractor can be a significant barrier to overcome.

There are specialized areas of technical expertise that a company may not want to do with in-house personnel. There may not be enough work on an ongoing basis to provide enough experience internally so that employees gain and maintain a necessary knowledge base. This can be true for both engineering and maintenance. If, for instance, the need to perform engineering calculations for relief system design occurs every three years it might be best to outsource this to a company which does this on a daily basis. If a large motor or gear box must be rebuilt every five years, the equipment manufacturer may have the most experienced and knowledgeable personnel.

Shared values begin with shared understanding of process safety. Both the operating and contracting company should be well-versed in process safety. This should be discussed in detail during the interview when a contractual arrangement is contemplated. In the interview process both should present:

- how they view and manage process safety elements
- how their personnel are trained in process safety
- what each company expects of the other

Failure to reach an agreement regarding these key areas is cause to veto the contract service agreement.

Like any other contract service, process safety requires clear reporting relationships. Process safety also requires the involvement of contract personnel. This is particularly true for design engineering and maintenance. Many process safety tasks such as process and equipment design, management of maintenance work orders, management of change, and involvement in pre-startup safety reviews (PSSRs) and hazard identifications require engineering and maintenance involvement and active participation. Issues such as coemployment concerns and liability need to be contractually resolved prior to implementation so that participation is expected on the day services begin. Both companies should communicate this to their respective personnel along with the message that a "them and us" attitude, especially as it relates to process safety, is unacceptable.

Training can also be a problem area. Most companies only provide basic hazard communication (HAZCOM) training to contractor employees and then rely on the contract company to provide any job-specific training to their own employees. In many instances this leaves the contract employees with gaps in their training, especially when it comes to unique issues and concerns related to the process or chemicals handled at the facility. When outsourcing functions which impact process safety, the details of the training provided to the contract employees should be carefully reviewed and customized or enhanced as needed.

Insourcing or outsourcing can be made to work effectively. Like other organizational structure changes this one requires effective planning, evaluation, and implementation.

7.7 STAFFING-LEVEL POLICY CHANGES (SHUTDOWNS, TURNAROUNDS, STARTUPS)

The associated activities of shutdowns, turnarounds, and startups present special challenges for process safety. Companies typically have special provisions or policies for additional staffing for these critical time periods. Changes to these special staffing policies should be considered as an OCM activity.

During shutdowns and startups, nonroutine operations and work are being performed. Shutdowns and startups are dynamic events as opposed to the more steady state event of continuous operations. This means the operating parameters of the production facility are in a state of flux. Control systems which may perform adequately under steady state may be challenged by a startup or shutdown. Indeed, startups may involve process controls which were installed in the associated turnaround and are operating for the first time in a real life situation.

The work that the staff is doing is also in a dynamic state. It generally requires more work by operations, unit management, and technical support to handle these events. Additionally it is nonroutine work, some of which may be being done for the first time.

Maintenance and construction work during a shutdown involves work which generally is seldom done, hence the reason for the shutdown. New construction and modifications generally involve work being done for the first time. Added to this mix are generally a lot of new people. Additional maintenance and construction personnel, design engineers, new equipment technical representatives, and others not normally in the plant add to the challenge.

Given this potent mix, it is little wonder that so many process incidents occur during shutdowns, turnarounds, and startups. To address these challenging situations, companies frequently have policies specifically aimed at staffing levels during startups and shutdowns. Any time these special policies are modified, special consideration should be given to their potential impact on process safety.

Tabletop or simulation exercises may be a good way to evaluate changes to these staffing policies. A procedure for assessing minimum control room staffing can be found in Example B.5 within Appendix B.

7.8 SPECIAL TRAINING REQUIREMENTS

For many of the organizational changes that have been discussed in this book, the training requirements pertain to training individuals to adequately handle new responsibilities. These training considerations are covered in Chapter 2 of this book. For the types of changes included in this chapter, the training pertains more to the decision makers who are writing the policies or making the policy changes.

In many cases, these types of policy changes are implemented by people situated very high in the organization who typically have many objectives to consider. In some cases, these decision makers may not be familiar with process safety and its implications for the business. In this case, it may be necessary to educate these high-level executives on what process safety is and how it may affect the business metrics that they may be more familiar with such as profitability and corporate image. One tool for this information sharing is a video and associated reference material that is being produced by CCPS entitled, "Executive Process Safety Seminar."

7.9 CONCLUSION

Policy changes should be carried out with process safety implications clearly understood. Once understood, those making the change may need to adjust the policy so that instead of being a detriment to process safety, it is a benefit. With any policy change, the stakeholders should clearly understand what is being changed and how it affects them. This requires good communications: first to state the policy change, second to receive feedback that it is understood, accepted, and implemented, and third to identify any shortcomings in the policy or its implementation.

REFERENCES

Baker Panel, *A Case Study for Review of BP's Process Safety Management Systems*, 2007.

http://www2.dupont.com/Heritage/en_US/1805_dupont/1805_in_depth.html.

APPENDIX A
EXAMPLE TOOLS FOR EVALUATING ORGANIZATIONAL CHANGES

There are a variety of tools which can be used to evaluate the impacts and risks associated with changes to an organization. Some of these tools have already been discussed or mentioned in the preceding chapters. Some of the tools are included here in their entirety and others are just references since they are publicly available documents. Feel free to adapt these tools to meet your specific needs.

A.1 WHAT-IF ANALYSIS

This is frequently the most commonly used tool for conducting risk assessments associated with organizational change management. Many companies just use a basic brainstorming approach to develop the necessary What-If questions used to evaluate the change. However, Tables A.1 – A.13 show one company's prepopulated What-If assessment. Each table addresses a different aspect of process safety for the organization.

Another type of What-If assessment utilizes a set of guide words to facilitate the brainstorming of relevant questions. An example set of such guide words is shown below:

- Competence
- Communication
- Experience
- Knowledge
- Workload

- Stress
- Quality of product
- Integrity of product
- Integrity of records
- Budget
- Staff appraisal
- Authorization level
- Authority
- Workforce reaction
- Morale
- Logistics
- Hours of overtime
- Location

A.2 CHECKLISTS

Another common option for evaluating organizational changes is to utilize a checklist. There are a variety of checklists available for evaluating organizational changes. They range from simple screening tools to more detailed risk assessments designed for developing and assigning action items. Several of these checklists are reviewed here.

The "Organisational Change and Major Accident Hazards, Chemical Information Sheet No. CHIS7," printed and published by the U.K. Health and Safety Executive 2003, includes a workflow for executing organizational change and contains checklists for each step of the workflow. These checklists include Getting Organized, Risk Assessment, Contractors, Risk Alert, Competence, and Transition. These checklists provide good information regarding the "dos" and "don'ts" for each step of the change process.

The Canadian Society for Chemical Engineering provides a good screening checklist for organizational changes to determine whether there is a significant safety impact for a particular type of organizational change. It provides guidance for which types of changes warrant a more detailed risk assessment. This screening checklist is included as Appendix 1 in "Managing the Health and Safety Impacts of Organizational Change."

The Canadian Society of Chemical Engineering's "Managing the Health and Safety Impacts of Organizational Change" also provides a checklist for use during a detailed risk assessment. It is organized similarly to a PHA or HAZID worksheet in that each line item includes a place to record the effect, a priority ranking (H/M/L), necessary safeguards, and action items. This checklist is included as Appendix 2 in the referenced document.

The Chemical Manufacturer's Association included several checklists in its document, "Management of Safety and Health during Organizational Change: A Resource and Tool Kit for Organizations Facing Change." In Attachment B of that document, there are several checklists to be used for evaluating changes. There are some simple screening checklists which allow the reviewer(s) to capture potential safety impacts and prioritize the issues. There are also several checklists, such as Safety and Health Management, Safe Work Practices, and Contractor Safety, geared towards identifying items that are impacted by the change, the possible effect, action items needed to maintain or improve safety, and assigning responsible parties and due dates for these actions.

Table A.14 shows one company's checklist, which can be used to facilitate a discussion of the impact and risks associated with a change. This doesn't directly highlight concerns, but it provides a list of topics which should be considered for possible concerns related to the change.

Another company utilized a more extensive set of checklists with specific issues that might be affected by an organizational change. Many of the items in these checklists identify various documents which might need to be updated as a result of the change. These checklists include columns to document the possible effects of the change, recommendations, and action item assignments.

These example checklists are included in Example B.2 in Appendix B of this book.

Another company put together a Microsoft™ Excel-based checklist of key issues to consider for possible impacts. If any of the questions are answered "Yes," then a MOC is required for the change. The checklist also provides space for recording further details regarding the impact related to that key question. This checklist is shown in Table A.15. Closely associated with this checklist is the task list shown in Table A.16. This task list can be used to flag what types of tasks are impacted, and what actions need to be taken, to ensure the task is properly assigned. This task list can be used as part of a task mapping or time study analysis which could be part of the risk assessment for the change.

TABLE A.1 Management Leadership, Commitment, and Accountability

MANAGEMENT REORGANIZATION/REDESIGN "WHAT IF?"/HAZARD IDENTIFICATION REVIEW
Notes: "What If"/HAZID team to be composed of a cross-section of affected personnel. Additional specifics to the reorganization/redesign may be added and documented. Team to identify findings and recommendations and if possible position/person responsible to address actions. Documentation of actions to be taken to be communicated with all affected personnel.
Management Leadership, Commitment, and Accountability
Management provides the perspective, establishes a system, sets the expectations, and provides the resources for successful operations. Assurance of operations integrity requires management leadership and commitment to be visible to the organization and accountability at all levels.

#	Issue/Expectation	Finding	Recommendation	Action
	A system for operations integrity management is established, communicated, and supported at every level in the organization			
	Management demonstrates commitment through active and visible participation in the operations integrity process			
	Operating management establishes the scope, priority, and pace for system implementation taking into account the complexity and risks involved in their operations			
	Line management has clear responsibility: roles, authorities, and accountabilities within the system are known and exercised. Access to specialist support is maintained			
	Clear goals and objectives are established for each element of the operations integrity system and performance is measured and evaluated			

TABLE A.1 Management Leadership, Commitment, and Accountability *(Continued)*

#	Issue/Expectation	Finding	Recommendation	Action
	Systems are in place to ensure that expectations are translated into procedures and practices			
	Systems are in place which ensure active employee involvement and participation in the operations integrity process. Learning is shared across the organization			
	A system is in place to assess performance and the degree of which expectations are met. The results are reported to senior management			
	How are the changes in staff and responsibilities communicated?			
	Will any safety issues fall off the plate by devoting time to organizing?			
	Does the existing organization provide any safety-related tasks or functions now that will be eliminated or reassigned by the new organization? Is there a system in place to ensure these tasks will continue to be performed after redesign?			
	How will different departments communicate? How does this affect communication with other departments? How do different shifts communicate the job status?			
	Are interfaces between different organizations recognized?			
	How will differences between departments/organizations be resolved?			

TABLE A.1 Management Leadership, Commitment, and Accountability *(Continued)*

#	Issue/Expectation	Finding	Recommendation	Action
	Do personnel know when to seek assistance? Are personnel encouraged to ask supervisors for assistance?			
	What system will be put in place to notify upper management of deficiencies?			
	Are individual responsibilities clearly defined? How do these relate to team responsibilities? How is performance monitored and measured? Do the workers themselves revise the roles and responsibilities? How often?			
	Were communication and teamwork considered in the redesign?			
	Is there adequate supervision?			
	What safety objectives are/will be established and how is attainment monitored?			
	Are the safety roles clearly understood by all (company and/or contract)? How is it verified?			
	Is a complete set of new roles and responsibilities available? How?			
	Will there be any additional responsibilities/workload after redesign?			
	How will daily tool box meetings be delivered?			
	Does the safety walkthrough program need modification to fit the new organization?			
	How will field-wide standards be applied and met?			
	How are problems with worker fatigue and/or stress resolved?			
	Will training requirements be affected? Are they sufficient?			

TABLE A.1 Management Leadership, Commitment, and Accountability *(Continued)*

#	Issue/Expectation	Finding	Recommendation	Action
	What is the supervisors' role in detecting and correcting human errors?			
	How do supervisors interact with workers?			
	What has been done to reduce the likelihood and/or consequences of potential human errors in the performance of these jobs?			
	Is upper management's commitment to employee health and safety clear? What policy statements communicate this?			
	Have supervisors and workers been told to err on the "safe side" whenever there is conflict between safety and production? Is safety the priority?			
	Is there a policy that clearly establishes who has the authority to stop work if safety requirements are not met? Is there a belief in the policy?			
	Is management of worker health and safety an essential part of the manager's daily activities?			
	How are managers held accountable? Will the tracking systems be revised to properly affect the reorganization?			
	Is health and safety regularly discussed in management meetings at all levels? Does this involve more than review of statistics?			
	How can management be assured that this redesign will not become a root cause for an incident?			
	Other management leadership, commitment, accountability issues?			

TABLE A.2 Risk Assessment and Management

Risk Assessment and Management				
Comprehensive risk assessments reduce risk and mitigate the consequences of operational, health, safety, and environmental incidents by providing essential information for making decisions.				
#	**Issue/Expectation**	**Finding**	**Recommendation**	**Action**
	A system is in place to identify sources of risk and hazards, assess their consequences and probabilities, and evaluate prevention and mitigation measures. Management of risk is a continuous process			
	Risk assessments are conducted for existing operations and new projects/modifications which address potential hazards and risks to personnel, facilities, the public, and the environment			
	Periodic risk assessments are performed by qualified personnel including, where appropriate, expertise from outside the immediate unit			
	Procedures are established to update risk assessments at specified intervals and as changes are planned			
	Assessed risks are addressed by specified levels of management appropriate to the nature and magnitude of the risk. Decisions are clearly documented			
	Does this redesign pose additional risk to the employees?			
	How does this redesign affect safety-critical equipment; safety systems, shutdown systems, pressure safety valves (PSV's), etc.?			
	How will prejob safety checklist be utilized? By operators? By production support? By maintenance?			
	What safety-related tasks and/or responsibilities will change for the operators? For production support? For control room?			
	Other risk assessment and management issues?			

TABLE A.3 Facilities Design and Construction

Facilities Design and Construction				
Inherent safety is enhanced, and environmental and health risks minimized, by using sound standards, procedures, and management systems for facility design, construction, and startup activities.				
#	**Issue/Expectation**	**Finding**	**Recommendation**	**Action**
	Project management systems, procedures, and accountabilities are documented, well understood, and executed by qualified personnel			
	Criteria are established, and procedures are in place, for conducting and documenting health, safety, and environmental risk assessments at specific stages from concept through startup			
	Appropriate baseline environmental and health data are collected prior to any new operations, facilities, major modifications, or personnel postings			
	Approved design practices and standards are used in the design and construction of new or modified facilities, which:			
	– meet or exceed applicable regulatory requirements			
	– embody responsible requirements where regulations do not exist			
	– incorporate operational expertise and experience from previous projects			
	– utilize HAZOPs			
	– apply the best available health, safety, and environmental technologies that are commercially viable			

TABLE A.3 Facilities Design and Construction *(Continued)*

#	Issue/Expectation	Finding	Recommendation	Action
	Deviation from approved design practices and standards or the approved design is permitted only after review and approval by the designated authority, and the rationale for the decision is documented			
	Quality control and inspection systems are in place which ensure that facilities meet design specifications and that construction is in accordance with the applicable standards			
	A review is performed before startup and documented to confirm that:			
	– construction is in accordance with specification			
	– health, safety, environmental, emergency, operations, and maintenance procedures are in place and adequate			
	– risk management recommendations have been addressed and required actions taken			
	– there has been adequate training of personnel			
	Other facilities design and construction issues?			

TABLE A.4 Operations and Maintenance

Operations and Maintenance
Operation of facilities within established parameters is essential to control risk. This requires procedures, structured inspection and maintenance systems, reliable safety systems and control devices, clean and tidy facilities, and qualified personnel who execute these procedures and practices consistently.

#	Issue/Expectation	Finding	Recommendation	Action
	Clearly assigned operating responsibilities and periodic reviews ensure appropriate levels of manning and skills to maintain integrity of operations through intended life			
	A system is in place that ensures the development of necessary procedures for operating, maintenance, inspection, and corrosion control which involve the workforce. This is updated at specified intervals and when changes are made			
	Programs are in place to ensure facility components are maintained and controlled within specified operating envelopes and in accordance with regulatory requirements			
	A quality assurance program exists to ensure that equipment replacement or modification maintains operations integrity			
	Systems are in place for work permits, handovers, and plant isolations which incorporate checks and authorizations that are consistent with mechanical and operations risks			
	Protective systems and devices are identified, tested, and undergo preventive maintenance. Appropriate documentation is available			
	A system is in place which controls the temporary disarming or deactivation of protective systems and devices			

TABLE A.4 Operations and Maintenance *(Continued)*

#	Issue/Expectation	Finding	Recommendation	Action
	Simultaneous operations and interfaces are assessed and systems are in place to manage risks			
	Systems are in place to monitor and minimize environmental impact, reduce wastes and emissions, and promote efficient energy use			
	Systems are in place to provide for the safe handling of hazardous materials or wastes and to meet regulatory requirements			
	A system is in place to provide for proper abandonment of facilities			
	Risk-based systems are in place for medical support, including first aid, medical advice and treatment, medical evacuation, medical assessment and health surveillance, and rehabilitation and health promotion			
	How does this redesign affect safety-critical equipment; safety systems, shutdown systems, PSVs, etc.?			
	What will be left unmonitored or lack current level of attention under the new organization?			
	Will operators or production support be working alone where previously they were working with others?			
	How will the bleeding/venting of gas requirements be affected by this redesign?			
	What system will be put in place after redesign to ensure safety-critical PMs will be performed?			

TABLE A.4 Operations and Maintenance *(Continued)*

#	Issue/Expectation	Finding	Recommendation	Action
	Will this redesign affect the PM schedule to ensure the reliability of safety-critical equipment and instrumentation, shutdowns, fire and gas, etc.?			
	Will procedures change for: hot work, opening process lines, work on energized electrical equipment, blinding before maintenance, confined space, etc?			
	Is it clearly defined when a job order is to be written and when repair work can be completed with a verbal request? Permit versus verbal approval?			
	How will this redesign affect permitting?			
	Are there certain jobs for each craft that should not be attempted by a single person? How are they communicated and controlled?			
	Who will perform the procedure writing?			
	How will the pigging/chemical crew be impacted? Permitting? Overall interaction with organization?			
	How will this redesign affect permitting for the workers involved?			
	Who can authorize procedural change?			
	Other operations and maintenance issues?			

TABLE A.5 Management of Change

Management of Change				
Changes in operations, process fluids, chemicals, procedures, site standards, facilities, or personnel are evaluated and managed to ensure that operations, safety, health, and environmental risks arising from these changes remain at an acceptable level. Changes in laws and regulations are reflected in facilities and operating practices to ensure ongoing compliance.				
#	**Issue/Expectation**	**Finding**	**Recommendation**	**Action**
	A system is in place for the management of temporary and permanent changes			
	Procedures address:			
	– authority approving changes			
	– analysis of health, safety, and environmental implications			
	– acquisition of required permits			
	– documentation, which includes reason for change			
	– communication of potential consequences and required compensating measures			
	– time limitations			
	– training			
	A system is in place to ensure that temporary changes do not exceed initial authorization for scope or time without review and approval			
	A system is in place to identify changes in laws, regulations, health, and environmental conditions and to reflect those changes in the facilities and operations affected			
	Does MOC need revisions?			
	Has proper MOC been followed for the redesign?			

TABLE A.6 Information/Documentation

Information/Documentation				
Current information regarding the configuration and capabilities of processes and facilities, properties of materials handled, potential health, safety, and environmental hazards and regulatory requirements is essential to assess and manage risk.				
#	**Issue/Expectation**	**Finding**	**Recommendation**	**Action**
	A system is in place to ensure that drawings and other pertinent documentation necessary for safe, environmentally sound operations and the maintenance of facilities are identified, accessible, and current			
	Responsibility is clearly defined for maintaining a process and facilities information/documentation system			
	Materials properties and potential hazards and operations risks are identified, documented, and communicated openly			
	Applicable regulations, permits, codes, and workplace standards and practices are identified and conflicts are resolved. The resulting operating requirements are documented and communicated to those affected.			
	Pertinent records covering operations, maintenance, inspections, and facility changes are maintained			
	Is the required documentation clearly defined for work?			
	Other information and documentation issues?			

TABLE A.7 Personnel and Training

Personnel and Training
Control of operations depends on people. Maintaining operations that are healthy, safe, and environmentally sound and conform to laws and regulations requires the careful selection, placement, ongoing assessment, and proper training of employees.

#	Issue/Expectation	Finding	Recommendation	Action
	A system is in place for the selection and placement of qualified employees to meet specified job requirements			
	Criteria are in place to ensure that necessary levels of individual and collective experience and knowledge are maintained and are carefully considered when personnel changes are made			
	A system is in place to provide initial and ongoing training to meet job and legal requirements. This includes:			
	–mechanisms for assessing effectiveness			
	– training documentation			
	–demonstrated competence on the job			
	A system is in place for periodic refresher training which includes evaluation and improvement of the training given and assessments of employee knowledge and skills relative to job requirements			
	Systems are in place to assess, document, and provide feedback on employee performance			
	Is there a minimum management staffing requirement?			
	How can you maintain experience? Is there a succession plan?			

TABLE A.7 Personnel and Training *(Continued)*

#	Issue/Expectation	Finding	Recommendation	Action
	What additional training and responsibilities are required?			
	How are additional techs/operators called out during an emergency or routine manpower shortage? Who makes the decision? Who determines qualifications?			
	Are the individual responsibilities clearly defined? How do these relate to team responsibilities?			
	Is any additional training required for any personnel at any level due to this redesign?			
	How are maintenance personnel trained on new processes, equipment, and procedures?			
	How are retraining needs identified?			
	How are the necessary skills and tasks converted into appropriate criteria for worker selection based on physical abilities, aptitudes, experience, etc.			
	Have the critical jobs and tasks been identified for each craft?			
	How is training effectiveness assessed? Evaluated?			
	Is any additional training required for personnel due to this redesign?			
	What training is given to personnel changing jobs or taking additional responsibilities? Is any additional training required for personnel due to this redesign? How is training effectiveness assessed? Evaluated?			
	Is additional training needed for operations supervisors? For maintenance supervisors? For team leaders?			
	Do the required training matrices need revision to fit the new organization?			

TABLE A.7 Personnel and Training *(Continued)*

#	Issue/Expectation	Finding	Recommendation	Action
	What time is allowed for training?			
	Is there a risk of implementing this redesign without training necessary personnel on the process overview?			
	Is additional training needed for operations supervisors who will be supervising maintenance technicians? Personnel not directly involved in the reorganization?			
	Have the critical jobs and tasks been identified for each craft?			
	Are there certain jobs for each craft that should not be attempted by a single person? How are they communicated and controlled?			
	Are the safety roles clearly understood by all employees affected by this redesign?			
	Is a complete set of the new roles and responsibilities available to the workers? How?			
	What training is given to workers changing jobs or given additional responsibilities?			
	Other personnel and training issues?			

TABLE A.8 Third-Party Services

Third-Party Services				
Third-party employees working on the company's behalf have an impact on its operations and its reputation. It is essential that they perform in a manner that is consistent and compatible with company policies and business objectives.				
#	**Issue/Expectation**	**Finding**	**Recommendation**	**Action**
	A system is in place for the objective evaluation and selection of third-party services which includes an assessment of their capability to perform work in a safe, healthy, and environmentally sound manner			
	Requirements for third-party personnel and performance are defined, communicated, and include a system for self-monitoring and accountability			
	A system is in place to ensure the effective management of interfaces between organizations providing and receiving services			
	Monitoring systems are used to assess third-party performance, provide feedback, and ensure deficiencies are corrected			
	Other third-party service issues?			

TABLE A.9 Incident Investigation and Analysis

Incident Investigation and Analysis
Effective incident investigation, reporting, and follow-up are necessary to achieve improvement in health, safety, and environmental performance. They provide the opportunity to learn from reported incidents and to use the information to take corrective action and prevent recurrence.

#	Issue/Expectation	Finding	Recommendation	Action
	Systems are in place for reporting, investigating, analyzing, and documenting safety, health, and environmental incidents and significant near-misses			
	Serious incidents are reported immediately and investigated by a team that includes appropriate external representation			
	Procedures exist for near-misses and incidents which:			
	– provide for timely investigation			
	– identify root causes and contributing factors			
	– determine actions needed to reduce the risk of related incidents			
	– ensure that appropriate action is taken and documented			
	– use legal resources as appropriate			
	Findings are retained and periodically analyzed to determine where improvements to practices, standards, procedures, or management systems are warranted. These are used as a basis for improvement			

TABLE A.9 Incident Investigation and Analysis *(Continued)*

#	Issue/Expectation	Finding	Recommendation	Action
	Systems are in place to share lessons learned from incidents and near-misses, trends, and successful practices across the company. These interact with others as appropriate and benchmark best practices to facilitate improvements in performance			
	Does the incident investigation system need revision to provide accurate reporting in the new organization?			
	Root cause			
	Other incident investigation and analysis issues?			

TABLE A.10 Community Awareness and Emergency Preparedness

Community Awareness and Emergency Preparedness
Community awareness is essential to maintaining public confidence in the integrity of our operations. Emergency planning and preparedness are essential to ensure that, in the event of an accident, all necessary actions are taken for the protection of the public, environment, company personnel, and assets.

#	Issue/Expectation	Finding	Recommendation	Action
	Open communications exist with employees, contractors, regulators, and the public			
	A system is in place to ensure the recognition of, and the response to, the community's expectations and concerns about the operations			
	Emergency response plans are documented, accessible, and clearly communicated. The plans include:			
	– organizational structure, responsibilities, and authorities			
	– internal and external communications procedures			
	– procedures for accessing personnel and equipment resources			
	– procedures for interfacing with other company and community emergency response organizations			
	– a process for periodic updates			
	Equipment, facilities, and trained personnel needed for responding to emergencies are defined and readily available			
	A system providing for simulations and drills is in place which includes consideration and involvement of external communications			
	What immediate actions are required by operators in an emergency? How will this change?			

TABLE A.10 Community Awareness and Emergency Preparedness *(Continued)*

#	Issue/Expectation	Finding	Recommendation	Action
	Are there clear procedures during emergencies for communications between operators, engineers, construction, control room, operations support, maintenance, response teams, and management?			
	Are there clear procedures during emergencies for communications between workers, emergency response personnel, and management?			
	Is there established protocol of command between the process units and Production Superintendent?			
	Who is on-scene command?			
	Who will respond to emergencies? In field, plant, and IMT (Incident Management Team) level?			
	How will the SRT (Safety Response Team), ERT (Emergency Response Team), and MERT (Medical Emergency Response Team) be affected?			
	How will the IMT organization be affected? Will there be incident management training requirements?			
	How will this redesign change roles in an emergency? How will it be communicated?			
	How will process unit emergency shutdowns be affected by this redesign? Does this redesign pose any additional risk?			
	How will this redesign affect startup after an emergency shutdown of process units?			

TABLE A.10 Community Awareness and Emergency Preparedness *(Continued)*

#	Issue/Expectation	Finding	Recommendation	Action
	How are additional techs/operators called out during an emergency? Who makes the decision?			
	Will techs/operators be working alone during emergencies where previously they have worked in pairs or with others?			
	How will this redesign change emergency response roles? How will it be communicated? Are all emergency procedures practiced regularly?			
	Are there conflicting roles for an operator during an emergency?			
	How will the emergency response plan be affected by this redesign?			
	Have jobs been analyzed for both routine and emergency activities?			
	Other community awareness and emergency preparedness issues?			

TABLE A.11 Operations Integrity Assessment and Improvement

Operations Integrity Assessment and Improvement				
A process that measures performance relative to expectations is essential to improve operations integrity and maintain accountability.				
#	**Issue/Expectation**	**Finding**	**Recommendation**	**Action**
	Measurable goals are used to assess performance and to target continuous improvement			
	A process exists for self-assessment of progress toward operations integrity goals			
	Audits are conducted using expertise outside the immediate unit			
	The frequency and scope of assessments reflect the complexity of the operation, level of risk, and performance history			
	A system is in place to ensure the resolution of findings from assessments			
	The effectiveness of the assessment process is reviewed periodically and findings are used to make improvements			
	Other assessment and improvement issues?			

TABLE A.12 Process Safety

Process Safety				
Process safety is critical to avoid catastrophic events such as fires and explosions				
#	**Issue/Expectation**	**Finding**	**Recommendation**	**Action**
	Will this redesign affect the process safety implementation and effectiveness of the facility process safety program or specifically any of its elements as follows not already covered in the Health, Safety, and Environmental Management System?			
	– Process safety information			
	– Process hazards analysis			
	– Operating procedures			
	– Employee participation			
	– Training			
	– Contractors			
	– Pre-startup safety review			
	– Mechanical integrity			
	– Hot work (and other) permitting			
	– Management of change			
	– Incident investigation			
	– Emergency planning and response			
	– Compliance audits			
	– Trade secrets			
	Other PSM issues?			

TABLE A.13 Generic What If

<table>
<tr><td colspan="5">Are there any other safety concerns? Environmental concerns? Health concerns? Overall operational integrity and risk management concerns?</td></tr>
<tr><th>#</th><th>Issue/Expectation</th><th>Finding</th><th>Recommendation</th><th>Action</th></tr>
<tr><td></td><td></td><td></td><td></td><td></td></tr>
<tr><td></td><td></td><td></td><td></td><td></td></tr>
</table>

TABLE A.14 Example Checklist for Organizational Management of Change

Structure

1. **Emergency/Abnormal Situation Preparedness**
   - 1.1 Does the change impact the personnel assignments of the EMT/ERT?
   - 1.2 Does the change impact the roles and responsibilities of personnel in EMT/ERT?
2. **Legal or Regulatory Requirement**
   - 2.1 Does the change adhere to local labor laws regarding working hours?
   - 2.2 Does the change impact the required number of competent persons?
3. **Documentation**
   - 3.1 Does the change impact the custodian of critical procedures?
   - 3.2 Does the change require changing task force structure/responsible positions/position roles/personnel?
4. **Operational Performance**
   - 4.1 Production
     - 4.1.1 Does the change impact plant-monitoring activities? (plant performance monitoring, analyzing potential equipment failures, process monitoring by panels, etc.)
     - 4.1.2 Does the change affect routine plant operating activities due to increase in workload? (sampling, housekeeping, equipment/machinery operations, etc.)
   - 4.2 Plant Reliability and Integrity
     - 4.2.1 Does the change affect plant equipment basic care (EBC)?
     - 4.2.2 Does the change affect preventative maintenance (PM) of the plant? (e.g., Will changes in staff make it difficult to complete PMs?)
     - 4.2.3 Does the change affect the inspection program in the plant?
   - 4.3 Personnel Capability
     - 4.3.1 Does the change require specific technical capabilities? (years of experience, certification, specific technical skills, etc.)
5. **Reporting/Layers of Accountability**
   - 5.1 Does the change result in an increase/decrease in organizational layers? (e.g., increase in reporting layers may result in disruption of workflow, increased red tape, and a decrease in reporting)
   - 5.2 Does the change result in overlapping of personnel reporting? (e.g., "dotted line" reporting)
   - 5.3 Does the change result in preferential work implementation? (e.g., combining two positions which are distinct can result in the position holder prioritizing the work scope differently)
6. **Communication**
   - 6.1 Does the change affect transfer/cascading of critical information? (e.g., information/communicating critical instructions or reporting will be affected)
7. **Transition**
   - 7.1 Does the change result in staff grouses/grievances? (worker strike, infringing employment contract agreement, etc.)

TABLE A.14 Example Checklist for Organizational Management of Change *(Continued)*

7.2 Does the change result in a transition period which will affect communication? (e.g., a new reporting structure or reporting personnel may affect flow of information as people are not familiar)

8. Employee Well-Being

8.1 Does the change affect staff well-being? (health due to additional exposure to hazardous conditions, physical and/or psychological effects, etc.)

9. Logistics

9.1 Does the change result in a space constraint? (office space, work stations, panel stations, fire assembly points, escape routes, etc.)

9.2 Does the change affect adequacy of staff amenities? (toilets, parking spaces, locker rooms, etc.)

9.3 Does the change affect adequacy of emergency safety equipment? (escape packs, breathing apparatus, etc.)

Personnel

1. Safety-Critical Responsibilities

1.1 Does the change impact plant emergency response capability? [e.g., EMT and ERT]

1.2 Legal or Regulatory Requirements

1.2.1 Does the change affect required certifications? (Stationary Engineer, Professional Engineer, etc.)

1.2.2 Does the change impact signatory requirement for any legal documents? (e.g., signatories for risk insurance)

1.3 Does the change affect the staffing for PHA leaders? (HAZOP, what-if, LOPA, etc.)

2. Safe Plant Operations

2.1 Does the change impact shift manning or maintenance manpower requirements?

2.2 Does the change impact EHS procedures? (lockout/tagout, hot work, confined space entry, etc.)

2.3 Does the change affect the plant's technical capability? (technical staff, trainers, SME, design reviewer, etc.)

2.4 Does the change affect on-going specific tasks which have significant EHS/process safety impact? (Note: These are special tasks/projects not normally specified in the position job descriptions.)

3. Assignment Tasks (Scheduled as To Do, Ongoing Tasks)

3.1 Does the change affect MOC change implementer?

3.2 Does the change affect authoring operating, maintenance, and EHS procedures? (Note: Especially when the incumbent is SME on the work/procedures he/she is authoring?)

3.3 Does the change affect closure of action items from critical EHS action database?

3.4 Does the change affect EHS initiatives? (EHS steering committee, emergency response plan (ERP) committee, etc.)

3.5 Other, please specify.

4. Knowledge Management

4.1 Does the responsible person have required relevant knowledge and experience which is not documented? (Note: Tacit knowledge or experience from the incumbent.)

TABLE A.15 Example MOC Organizational Change Checklist

MOC Organizational Change

Site: ____________

Start Date: ____________

MOC Name: ____________ End Date: ____________

Facilitator: ____________ MOC #: ____________

Recorder: ____________ Originator: ____________

Step	MOC Organizational Changes	Requires MOC	No MOC required	What is impacted?	How impacted?	Consequence of impact?	How to address impact?	Is impact item fully covered after the change to maintain critical EHS function or compliance?
1	Normal MOC procedure of gathering backup information and a representative cross section of plant personnel including Human Relations and EHS to review and approve the change.							
2	Describe the existing organization or position and then describe the proposed new organization or position and associated responsibilities.							
3	Ask the following:							

TABLE A.15 Example MOC Organizational Change Checklist *(Continued)*

Step	MOC Organizational Changes	Requires MOC	No MOC required	What is impacted?	How impacted?	Consequence of impact?	How to address impact?	Is impact item fully covered after the change to maintain critical EHS function or compliance?
a	**Emergency Response**							
	Will this change affect the ability of personnel to respond safely during an emergency plant upset?	Yes	No					
	Will this change affect the response to those plans?	Yes	No					
	Are the requirements of 29 CFR 1910.38(a) impacted negatively?	Yes	No					
	Is the site ERP impacted?	Yes	No					
	How?							
b	**Process Safety Management**							
	Does the change affect the ability to assure PSM is maintained if the unit is a covered process?	Yes	No					
	Will mechanical integrity testing or inspections be impacted?	Yes	No					
	Will the unit still have proper staffing to conduct PHAs?	Yes	No					

Step	MOC Organizational Changes	Requires MOC	No MOC required	What is impacted?	How impacted?	Consequence of impact?	How to address impact?	Is impact item fully covered after the change to maintain critical EHS function or compliance?
	Will documentation of process safety information (PSI) be impacted?	Yes	No					
	Is there a training plan/schedule in place for employees new to the area?	Yes	No					
	Will annual review of SOPs or training be impacted?	Yes	No					
	Are there still available personnel trained in incident investigation per policy 2.01?	Yes	No					
	Are there any other concerns about maintaining PSM?	Yes	No					
c	**Safety**							
	Will the overall safety of the plant or personnel be impacted?	Yes	No					
	Will scheduled safety inspections of items such as safety showers, fire extinguishers, hoists, ladders, etc., be impacted?	Yes	No					

TABLE A.15 Example MOC Organizational Change Checklist *(Continued)*

Step	MOC Organizational Changes	Requires MOC	No MOC required	What is impacted?	How impacted?	Consequence of impact?	How to address impact?	Is impact item fully covered after the change to maintain critical EHS function or compliance?
	Will completion of safety permits be impacted?	Yes	No					
	Will safety meetings be impacted or reduced?	Yes	No					
	Will safety contact programs be impacted?	Yes	No					
	Does the change involve a person(s) responsible for conducting safety audits or reviews?	Yes	No					
d	**Environmental**							
	Will the change impact a person(s) position who is responsible for environmental tasks such as sampling, permitting, leak detecting, recordkeeping, inspections, etc., be impacted?	Yes	No					
	Will waste disposal, housekeeping, inspections, etc., be affected?	Yes	No					
	Will change impact the ability to determine permit violations?	Yes	No					

Step	MOC Organizational Changes	Requires MOC	No MOC required	What is impacted?	How impacted?	Consequence of impact?	How to address impact?	Is impact item fully covered after the change to maintain critical EHS function or compliance?
e	**Industrial Hygiene, Health**							
	Will the change affect the ability to perform monitoring or gathering of IH exposure data?	Yes	No					
	Will schedules of data gathering be affected?	Yes	No					
	Is health monitoring affected?	Yes	No					
f	**EHS Training**							
	Will training in EHS topics be impacted?	Yes	No					
	What specifically will be impacted?	Yes	No					
	Will compliance- or management-required training be changed or modified as a result of this organization change?	Yes	No					
	How?							
	Will the schedule for training be modified?	Yes	No					
	How?							

TABLE A.15 Example MOC Organizational Change Checklist *(Continued)*

Step	MOC Organizational Changes	Requires MOC	No MOC required	What is impacted?	How impacted?	Consequence of impact?	How to address impact?	Is impact item fully covered after the change to maintain critical EHS function or compliance?
	Will training for new or relocated personnel be impacted?	Yes	No					
	How?							
g	**Other EHS**							
	Are there any other compliance or management requirements which would be affected by this change?	Yes	No					
	What?							
	Will this change impact a previous recommendation from a PHA?	Yes	No					
	Are procedures in place to prevent potential workplace violence/sabotage?	Yes	No					
	TIME STUDY - Conduct a time/job task study for changes which will affect personnel's work load to determine if sufficient time and schedule are available to safely and fully complete the new or added tasks.							

TABLE A.16 Example Task List

Mark "X" by each task that is impacted by the personnel change. Then complete the remaining fields for each impacted task.				
X	**Category/Task**	**Impact?**	**Action to be Taken**	**Other Considerations**
PSM				
	Process Safety information			
	Process Hazard Analysis			
	Lead MOCs			
	Lead pre-startup safety reviews (PSSRs)			
	Operating Procedures			
	Training			
	Mechanical Integrity			
	Facilitate Incident Investigations			
	Other (list)			
	Other (list)			
EHS				
	Fit testing, mask training, enclosure training, etc.			
	Emergency planning and response			
	Equipment files			
	Safety/team meetings			
	Compliance permitting			
	Audits [e.g., safety, Resource Conservation and Recovery Act (RCRA), environmental]			
	EHS portal (e.g., data entry, validation, action item closure)			
	Training (e.g., RCRA, portal, spill reporting)			
	Waste management (e.g., manager, unit responsibilities)			
	Leak detection and repair (LDAR; e.g., inspections, records, invoicing)			
	Permit compliance (e.g., sampling, data entry, reporting)			
	RCRA Subpart BB			

TABLE A.16 Example Task List (*Continued*)

X	Category/Task	Impact?	Action to be Taken	Other Considerations
	Updated unit organizational chart			
	Spill prevention control, and countermeasures (SPCC; e.g., plan, inspection)			
	Superfund Amendments and Reauthorization Act (SARA ;e.g., tier two, toxic release, updates)			
	Inventory levels			
	New material introduction			
	Community emergency response team (CERT) transition plan			
	Continuous monitoring (e.g., emissions, opacity)			
	Ozone depleting substances			
	Other (list)			
Maintenance				
	MMS			
	I/E plan			
	Mechanic plan			
	Generator turnaround plan			
	Compressor rebuilds			
	Analyzer upgrades			
	Maintenance work area/bag changeout area			
	Plan for Fab area			
	Building door keys			
	Fencing plan			
	Scaffolding plan			
	Access plan			
	Track mobile plan			
	New air conditioning plan			
	Bobcat loader			
	Other (list)			
Operations				
	Staffing and coverage plan			
	Overall training plan			
	Log sheets			
	Daily production reporting			

TABLE A.16 Example Task List (*Continued*)

X	Category/Task	Impact?	Action to be Taken	Other Considerations
General				
	Work request system			
	DCS/automated information management (AIM) system connections			
	Outage planning			
	Quality chart			
	Stable operation			
	Safety audits			
	Site assessment realignment			
	Purchase orders			
	Suggestion program			
	Appraisals			
	Lab supplies			
	Lab testing			
	Work orders			
	Document retention			
	Other (list)			

A.3 OTHER RISK ASSESSMENT TOOLS

In addition to checklists, other methods may be used to evaluate the risks associated with organizational changes. Some specific examples of these other tools are discussed here and can be adapted as needed for specific organizational needs.

One tool that can be used to help visualize the hazards and results is the Bow Tie. This method is described in the article, "Identifying Key Safety Roles during Organizational Change." This article explains the methodology and goes through some examples of how it is used.

The Contract Research Report 348/2001, entitled "Assessing the Safety of Staffing Arrangements for Process Operations in the Chemical and Allied Industries," discusses a method specifically geared toward evaluating staffing requirements. A detailed example of how to use this tool is included in the next section. The assessment method includes a physical assessment along with a series of ladders to determine the minimum staffing levels for operations.

A.4 SPECIAL COMPETENCY ASSESSMENT FOR CONTROL ROOM STAFF

Research has indicated that control room operators believe organizational factors can have a bearing on their ability to avoid misinterpretation errors. They are able to manage their work to minimize distractions and optimize focus as long as they have a clear sense of their own capabilities and the demands on their time. Thoroughly understanding and accurately mapping all of the knowledge and experience required for each position are absolutely vital. When working conditions change, whether expectedly or unexpectedly, there is always a risk that some existing responsibilities will be neglected or overlooked. By establishing an initial basis for comparison and fully documenting the duties of each staff member at that point in time, these oversights can be avoided in critical situations in the future. There are a number of techniques available that can be used for establishing baseline competency requirements for current job descriptions. This section will focus on an assessment method which is particularly well suited to assessing the staffing levels within a control room or production unit. Additional tools can be found elsewhere in this appendix and in Appendix B.

When implementing any of these techniques for assessing staffing level, it is wise to repeatedly remind personnel that they are being used to improve procedures and evaluate the impacts of organizational changes, rather than as a tool for penalizing employees for gaps in knowledge or documentation. It is also imperative that these reminders be heeded by management at all times, lest the movement toward improved process safety be set back by the resultant breakdown in trust between personnel and management.

The Health and Safety Executive in the United Kingdom has outlined an assessment method for documenting work arrangements according to established rating scales. This method, detailed in Figure A.1, incorporates a number of different techniques useful in analyzing existing staffing levels and eliminating knowledge gaps in preparation for hazardous activities. It is designed to recognize when staffing numbers are too low to control an existing process, though it cannot be inferred that the method will suggest a minimum or optimal number of staff. With some effort the method could be adjusted to examine any of the changes discussed in this chapter.

One technique described by the Health and Safety Executive is to break down the tasks within a unit to their simplest activities. Existing activities should be observed and documented, including not only physical procedures but any associated communication that is necessary to complete those procedures. These observations can be incorporated into a matrix so that it becomes simple to recognize which tasks, information, or decisions are required for each unit team member to be able to complete his or her duties. This technique is intended for use only in evaluating routine daily operations and is not suitable for analyzing emergency or upset conditions.

One way to accomplish this is to assign an outside observer or group of observers to document regular activities throughout a typical shift. Several regular shifts may be observed to obtain the most complete documentation possible, or day and night shifts may be observed to compare requirements for each shift. The observers work among the employees to document procedures and tasks as they occur. They note any communication that is required to complete these procedures and tasks and are willing and able to ask for clarification from supervisors, unit operators, and technicians as needed. At some point after the shift, observers can meet with the employees to discuss their observations and obtain additional information that may be helpful for their evaluation.

This exercise should be fully explained to all personnel so that the flow of information is not hampered by concerns about perceived job performance or other factors. If this exercise is being performed to gather information and not as a means to justify a staff reduction, that should be made clear at the outset or the information gathered may be skewed. However, keep in mind that the very act of being observed may still cause some people to modify their behavior. Once the data are compiled, a matrix can be generated to reflect each position within the shift structure, its corresponding duties, and the information, approvals, or decisions required to complete those duties. Providing this matrix to the shift or shifts under observation would afford an opportunity for feedback or clarification of any misinterpreted observations. Including information from both the observer and the observed will guarantee the most vigorous evaluation of routine operations possible. If necessary, this technique may be scaled down by only requiring observation of those procedures during each shift that directly relate to potential process safety concerns, though it is difficult to ensure that all appropriate activities are identified. A detailed description of regular daily tasks would be required for each shift observed. The observation team would then be able to determine which tasks should be documented.

Another useful technique highlighted by the Health and Safety Executive is for each member of the unit to complete a self assessment of their daily activities. By using questionnaires and/or shift diaries, personnel can provide information on their own experiences throughout the day, including both positive and negative feedback. Questionnaires can be very specific regarding work demands, coping mechanisms, alertness, and other organizational factors. Diaries may be used in conjunction with a simulation exercise so that employees can record points in the exercise when they lacked information necessary to proceed or when existing protocols were helpful.

This technique can be incorporated into the regular training activities outlined above when reviewing startup, shutdown, or emergency procedures. In the weeks or months prior to the training exercise, each employee is asked to complete a survey regarding his or her normal duties. The survey should include open-ended questions in which the employee can express concerns about procedures or provide feedback on job aids that have been helpful in performing his or her daily tasks. The survey is not intended to be used as an examination of job responsibilities, but rather as a guide for management to focus on deficiencies in process documentation and instruction.

Each employee can then be asked to complete a job diary for one complete shift. This diary should outline daily responsibilities, providing as much detail as possible. Physical tasks or procedures ought to be documented, as well as all of the communication steps required to accomplish these tasks and procedures. A successful job diary should enable management to re-create an employee's daily routine from shift change to shift change. The questionnaire and accompanying shift log will be valuable tools in appraising the efficiency of the existing organizational structure during normal operations.

Once these two tools have been completed, training exercises can begin. During a process simulation, each employee is asked to complete an additional diary, documenting the procedures used during the exercise. The simulation should proceed at a pace that will enable personnel to review unfamiliar procedures, ask questions, and document practices and observations in their diaries. Again, these diaries can prove to be meaningful tools in developing an overall strategy for improving process safety. This technique can be used periodically to refresh training and to capture information months after new organizational changes have been implemented.

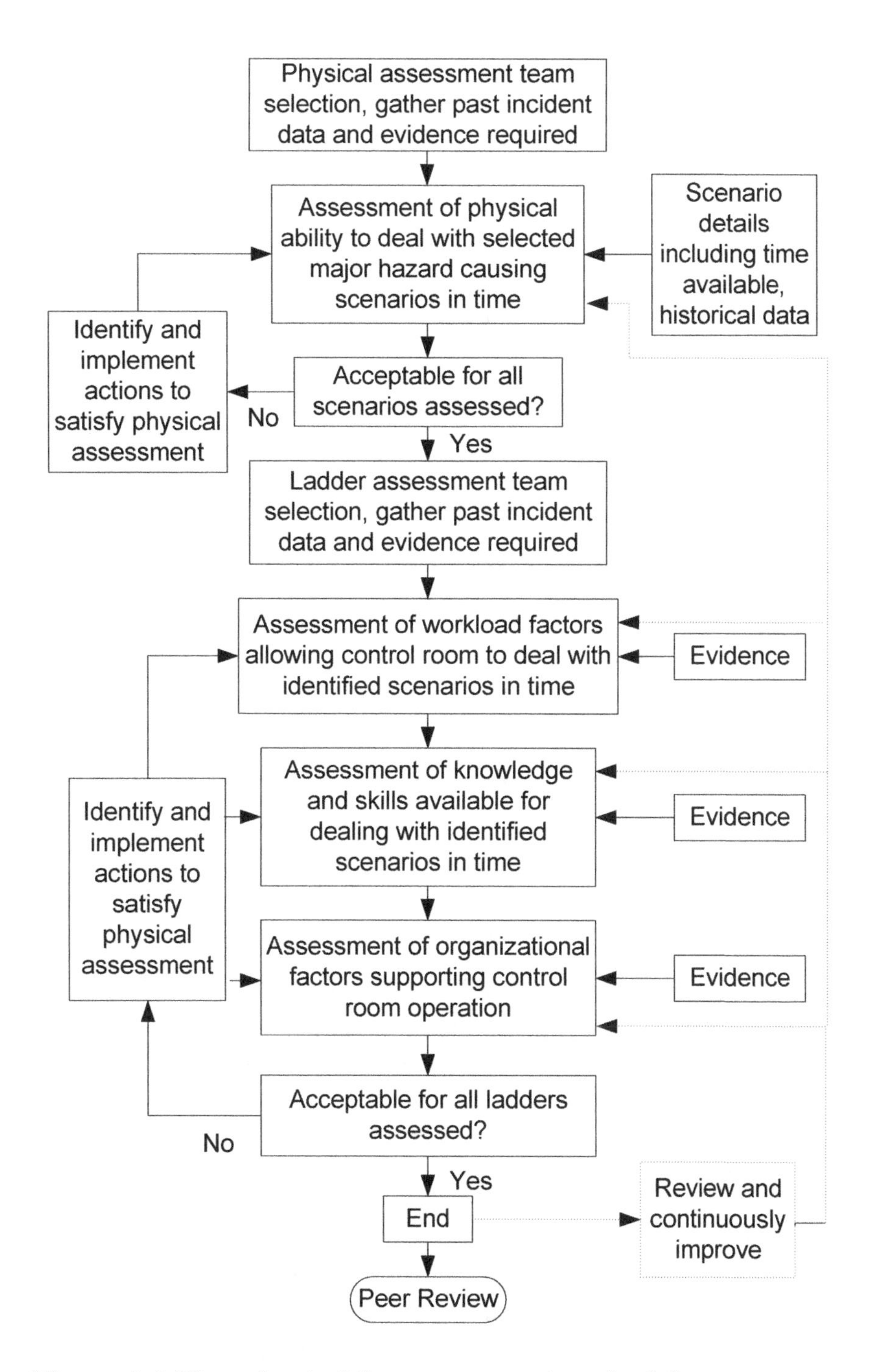

Figure A.1 Flow chart of the aassessment method for documenting work arrangement.

The assessment technique focused on by the Health and Safety Executive involves three phases of assessment of an existing organizational structure:

- Appraisal of technical factors
- Appraisal of individual factors
- Appraisal of organizational factors

Appraisal of technical factors focuses on the equipment and processes within the unit and is considered the physical assessment of current conditions. It is founded on techniques such as hazard and operability studies and fault tree analysis. This technique uses question trees to identify weaknesses in these areas. If weaknesses are identified during the question tree portion of the method, these concerns should be addressed before attempting the ladder assessment portions.

A representative tree from the physical assessment spotlights the alarm system within the control room. By initially asking whether the control room is continuously manned, it sets up a series of fault-tree-type questions. If the answer is "yes," then there is no further issue over whether an alarm will be recognized. If the answer is "no," questions develop regarding where the operator goes, how long he or she may be gone, and secondary methods by which the operator can be notified of an alarm while absent from the control room. Lack of a way to contact the operator in this situation, for example, would result in a "Fail" mark for this assessment. This deficiency should be corrected, and all other assessments completed to success, before moving on to evaluate the individual and organizational aspects of the existing structure.

Individual and organizational factors are each investigated separately using a total of 11 anchored descriptive rating scales called ladders. Seven ladders are used to diagnose individual factor issues and 4 ladders comprise the organizational factors evaluation. By asking a series of yes/no and open-ended questions of the employees involved in the change, various aspects of individual skills and the organization of the unit can be plotted on these ladders. The ladders compare the responses to these questions with predetermined criteria on an ascending scale from poor practices to best practices. Changes in location, shifts, or work hours are likely to directly affect the physical ability and workload elements of these ladders. Each ladder includes a horizontal dotted line. The lettering sequence also changes at this dotted line. If a response falls below this line, the practice in that particular area is considered unacceptable; if it falls above the line it is an acceptable practice. The ladders are used together to provide a detailed analysis of existing competencies within a unit and to determine where improvements should be considered.

One of the ladders used in this technique focuses on the management of operating procedures, as illustrated in Table A.17. For the facility to be within the acceptable range on this particular ladder, answers should clearly indicate that operators and management have a clear comprehension of routine operating procedures. Understandable and concise instructions, complete with checklists and job aids, should be readily available to those responsible for the daily operations of the area. Operators should demonstrate an awareness of who writes the procedures for their unit, when they are updated, and whether they are audited. Management is then questioned regarding specifics about how procedures are updated and approved and should provide information about the QC process for procedural changes. Documents on-site should include operating procedures with dates, authors, approvers, and revision numbers, a quality manual that specifies how procedures are managed, and any available results from audits of the procedures. All 11 rating ladders and additional details regarding their use are provided in the Health and Safety Executive reference, "Assessing the Safety of Staffing Arrangements for Process Operations in the Chemical and Allied Industries."

TABLE A.17 Ladder for Management Procedures

Grade	Description	Explanation of Progression	Rationale
A	Information about best practices is proactively shared between production units and sites	Further evidence of a learning organization keen to share its experience and learn from others	
B	The procedure quality control system is subject to review and continuous improvement	Evidence of a learning organization committed to continuous improvement	
C	There is a comprehensive procedure quality control system which the operations team is an integral part of and which ensures that procedures are recalled and updated when there is any process, equipment, or staff change which necessitates it	Procedures are managed and incorporated in the management of change process. The importance of operator involvement to encourage ownership and therefore active use of procedures is encouraged	
D	The operations team is responsible for ensuring that procedures are up to date and reflect current best practice	Further encourages ownership and should help to ensure procedures are updated and reflect best practice as it continuously improves	
E	Existing procedures are audited regularly to ensure they represent current best practice used by the operating teams	It is recognized that practices vary across shift teams and working practices may change with time. Auditing ensures unsafe or poor practices are identified and rectified, encourages continuous improvement, and ensures operating procedures are living, active documents	

TABLE A.17 Ladder for Management Procedures *(Continued)*

Grade	Description	Explanation of Progression	Rationale
F	Operators are part of the procedure-writing team and all operators are fully trained with new procedures and given the opportunity to provide feedback on the procedures before they are approved and made formal	It is recognized that operator involvement will increase ownership and reflect working practice. This will encourage active use and operator-generated updates	
G	The procedures are accessed close to point of use and are presented in a clear, concise manner with checklists and other job aids for critical operations	Procedures have been designed to be used and it has been recognized that providing checklists and other job aids will encourage use and help to reduce errors in application. They also prevent operators generating their own job aids which may become out of date as they are uncontrolled	
X	It is clear which procedure should be used for a particular task or situation. All information required for a particular task or operation is kept together and is easily referenced	No duplication of procedures, therefore no confusion about which is the correct version. Are acceptable and operators can rely on the accuracy of the information	
Y	New procedures are provided for significant process changes. There is a quick run-through given to operators when the procedures are introduced	No operator training for new procedures—even for significant process changes. Small process changes or changes in work practice are missed and procedures tend to "drift" out of date over time	
Z	Procedures were written several years ago and there have been few if any changes. There is no evidence of procedure quality control system. Operators play no part in the writing. There is a quick run-through of procedures given to operators when they are introduced	Poor practice, staffing arrangements do not fulfill any of the rungs above	

REFERENCES

Canadian Society for Chemical Engineering, *Managing the Health and Safety Impacts of Organizational Change*, Ontario, 2004.

Chemical Manufacturer's Association, *Management of Safety and Health During Organizational Change*, Washington, DC, 1998. This organization is currently known as the American Chemistry Council.

Davidson, P.A. and Mooney, S.D., *Identifying Key Safety Roles During Organizational Change*, Unilever, New York, 2009.

Entec UK LTD, Contract Research Report 348/2001, *Assessing the Safety of Staffing Arrangements for Process Operations in the Chemical and Allied Industries*, Shropshire, 2001.

Health and Safety Executive Information Sheet, *Organisational Change and Major Accident Hazards*, CHIS7, Castleford, 2003.

APPENDIX B
EXAMPLE PROCEDURES FOR MANAGING ORGANIZATIONAL CHANGES

Chapter 2 discusses the importance of having a procedure to manage organizational changes and provides a framework for such a procedure. This appendix includes some example procedures from various companies to provide some insight into different ways that organizational changes can be managed. Feel free to adapt these procedures as appropriate for your organization.

Example B.1 includes a procedure which differentiates between personnel changes and larger organizational changes. Example B.2 includes the detailed forms for this procedure which are related to organizational changes. It includes an extensive set of checklists with specific issues that might be affected by an organizational change. Example B.3 includes the simpler form for considering risks and action items associated with personnel changes.

Example B.4 is a basic procedure for managing organizational changes. It doesn't differentiate between personnel changes and other types of organizational changes. It suggests using guide words and brainstorming for the risk assessment.

Example B.5 is a procedure focused on assessing the minimum staffing levels for a process unit. This is intended to be a supplement to a more traditional MOC procedure. This procedure focuses on specific emergency procedures and determines the number of people required to get the operation to a safe state in a reasonable period of time. They start with a table-top review of the procedure, then consider all necessary steps and lay out parallel and sequential sets of activities. Consideration is also given to the locations for the various tasks and any potential hindrances which may occur in carrying out those tasks in determining the minimum staffing levels.

Additional information about what should be included in an OCM procedure can be found in the Chemical Manufacturer's Association's "Management of Safety and Health during Organizational Change" paper and in the Canadian Society for Chemical Engineering's "Managing the Health and Safety Impacts of Organizational Change" article.

Example B.1 MOC Procedure for Organizational and Personnel Changes

1.0 PURPOSE

The purpose of this standard is to ensure a systematic method to evaluate proposed organizational and personnel changes within XYZ facility operations, emergency response, process safety and mechanical integrity, and EHS responsibilities. The objective is to identify and evaluate the risks to the business arising from the change and then implement the steps necessary to mitigate those risks.

2.0 SCOPE

This standard applies to XYZ operations. Organizational changes require a more formal risk-based, controlled decision-making process. Personnel changes can be managed with a simple and "thought-based" approach.

This required standard does not replace existing processes for staffing and selection of qualified candidates to fill positions in the XYZ organization but rather supplements the staffing and selection process. For organizational changes, the management of organizational change (MOOC) is conducted prior to or concurrent with staffing and selection. For personnel changes, the management of personnel change (MOPC) is conducted after staffing and selection.

Not all positions are included in the scope of this required standard. Only positions with job responsibilities for operations, emergency response, process safety and mechanical integrity, and EHS are included.

3.0 DEFINITIONS

3.1 Organizational Change (MOOC) – a modification to existing organizational structure, reporting relationships, or permanent staffing levels in the operations, emergency response, process safety and mechanical integrity, and EHS functions.

- Organizational changes include consolidation of departments and changes in the number or designation of process units or assets.
- Organizational changes include reductions/ increases in permanent staffing levels and moving job duties from one department to another.

3.2 Personnel Change (MOPC) – the movement of individual personnel into or out of an existing position or new responsibilities requiring new skills and competencies assigned to an existing position.

- Personnel changes include the facility manager and all other positions in the functions of operations, emergency response, process safety and mechanical integrity, and EHS.
- Personnel changes include moves within the facility as well as experienced transfers and new hires from outside the facility.
- Production operator and craft worker positions which are managed under a formal qualification and training program are **not** required to have an associated MOPC. However, the qualification and training program must be defined,

documented, and training records appropriately retained to demonstrate proper management of these personnel changes when moving personnel into or out of existing positions or adding new responsibilities requiring new skills and competencies.

4.0 RESPONSIBILITIES

4.1 Each facility shall develop a site procedure that:

4.1.1 Defines the job positions which fall under the scope of this required standard, either by negative exception or positive declaration.

4.1.2 Designates key facility management personnel the responsibility for identifying organizational and personnel changes required by this standard, ensuring the risks arising from those changes are evaluated and the actions necessary to mitigate those risks are implemented.

- For organizational changes this shall be the operations manager, maintenance manager, technical manager, and/or EHS manager(s).
- For personnel changes this shall be the supervisor.

4.1.3 Identifies authority requirements for approving organizational (MOOC) and personnel (MOPC) reviews and actions identified from the reviews.

- All organizational changes shall be approved by the facility manager or documented delegate prior to or concurrent with implementation.
- Personnel changes may be approved by the supervisor and can occur after the

change effective date. In the case of a personnel change of the facility manager position (for which no on-site supervisor is present); the operations, maintenance, technical, or EHS manager shall be designated to review and approve the change.

4.1.4 Ensures actions identified by the MOOC or MOPC are implemented and documentation is retained.

4.1.5 Includes reference to the operator and craft worker qualification and training programs.

4.2 XYZ EHS – Periodically review and update this document as necessary to address deficiencies and emerging issues.

5.0 PROCEDURE REQUIREMENTS

5.1 Change Identification – Organizational and personnel changes required by this standard shall be identified and documented. Site procedures will define who is responsible for ensuring changes are identified and documented. This responsibility shall be designated to key management personnel (i.e., operations, maintenance, technical, or EHS Manager for organizational changes and the supervisor for personnel changes).

5.2 Change Review – Changes will be reviewed by appropriate personnel, depending on the change complexity and criticality.

5.2.1 Personnel changes may be reviewed and documented by the supervisor only. In the case of a personnel change of the facility manager position (for which no onsite supervisor is present), the operations, maintenance, technical or EHS manager shall be designated to review and approve the change.

5.2.2 Organizational changes shall be evaluated and documented by a MOOC team comprised of the appropriate mix of experience and skills to properly assess the change. MOOC teams shall have a minimum of two members but for complex changes will be comprised of membership similar to a PHA team. Factors considered in the formation of the team include experience, understanding of human factors, technical capability, organizational and business expertise, and knowledge of risk management principles.

5.3 Change Assessment – Each position that is being changed, moved, or eliminated shall be assessed to ensure operations, emergency response, process safety and mechanical integrity, and EHS responsibilities are maintained.

5.3.1 For personnel changes, a thought-based approach using a simple checklist is sufficient to assess the change. Reference the Optional Template – *MOPC Review and Approval Form* referenced in this required standard as an example that may be used to document a MOPC.

5.3.2 For organizational changes, each position that is being changed, moved, or eliminated shall be assessed and key responsibilities mapped to a new or already existing position. The MOOC team will ensure the proper planning of the change by identifying and documenting any actions needed to maintain or improve the site's safety. The health and safety impacts of the following areas shall be considered as they apply to the change:

- Operations and Safety Effectiveness
- Safety and Health Management
- Safe Work Practices
- Process Safety Management
- Contractor Safety

- Emergency Response
- Safety and Health Regulatory Compliance
- Occupational Health
- Process Unit Operability and Safety Effectiveness

Reference the Optional Template – *MOOC Review and Approval Form* referenced in this required standard as an example that may be used to document a MOOC. Very complex changes may require a more in depth assessment and such documentation may be used in lieu of this MOOC template. In such cases, the team conducts the hazard review functioning similarly to a PHA team. The documentation from such reviews is processed with the change request.

5.4 Change Approval – All organizational (MOOC) changes shall be approved by the facility manager or documented delegate prior to implementation. Personnel changes may be approved by the supervisor and this approval can happen after the change effective date.

5.5 Action Identification and Closure – Actions needed to properly manage the change shall be identified, responsible parties assigned with due dates, and action items tracked to closure and documented.

5.6 Documentation – Completed MOOC and MOPC documentation shall be maintained in accordance with process safety management, management of change, and records management policies.

5.7 Communication – All approved changes shall be communicated to the affected workforce at the site.

Example B.2 Template MOOC Review and Approval Form

I approve the Management of Organizational Change plan defined below (or attached in lieu of). All action items have been assigned and are being tracked within a management system to ensure completion.

Facility Manager Signature:______________________________

MOOC Title (Title of the organizational change and the effective date.) NOTE: Often the effective date is phased over a period of time.
MOOC Coordinator (Person who is responsible for coordinating the MOOC review and documentation including documentation of completed action items. This person is typically assigned by facility management.)
Technical Basis of Change (Document here why the change is being made.)
MOOC Team (List here each team member and their function/job title.)
INSTRUCTIONS – Either using the checklists in Tables B.1 through B.9 or by attaching the documentation from another appropriate hazard review methodology, review the organizational change to ensure the ongoing effectiveness of the following areas: Operations and Safety Effectiveness, Safety and Health Management, Safe Work Practices, Process Safety Management, Contractor Safety, Emergency Response, EHS Regulatory Compliance, Occupational Health, Process Unit Operability, and Safety Effectiveness.

Date:____

TABLE B.1 Operations and Safety Effectiveness Checklist

1 – OPERATIONS AND SAFETY EFFECTIVENESS CHECKLIST					
Factor or Issue of Concern	**Yes/ No/ NA**	**Possible Effect**	**Action to Maintain or Improve Safety**	**Action For**	**By (Date)**
Planning and Participation					
Has a process been established to involve affected personnel in the safety and health considerations pertaining to the change?					
Administrative Factors					
Are new lines of authority established and understood by all, particularly with regard to around-the-clock communications?					
Are the procedures in place for decision making, particularly in off-hours?					
Have the procedures for calling out additional personnel been addressed?					
Have the affected employees been communicated as to the position change and how it affects their job responsibility?					
Is/are the personnel in new assignments qualified for the position?					

TABLE B.2 Safety and Health Management Checklist

2 – SAFETY AND HEALTH MANAGEMENT CHECKLIST					
Could the Change....	**Yes/ No/ NA**	**Possible Effect**	**Action to Maintain or Improve Safety**	**Action For**	**By (Date)**
Affect the make-up of the plant's emergency response teams or fire brigade?					
Affect the perception of line management commitment to safety?					
Affect the visibility of line management in the plant (i.e., "walking the talk?")					
Affect accountability for safety?					
Require revision or consideration of the EHS policy statement?					
Require revision of the Process Hazard Overview booklet for the facility?					
Require revision to safety awareness programs?					
Require changes in the manner in which departmental safety meetings are performed?					
Require changes in the manner in which safety bulletins or periodic newsletters are prepared or issued?					
Require changes in the way that safety and health information is communicated?					
Affect the injury and illness reporting and investigation procedure?					

TABLE B.2 Safety and Health Management Checklist *(Continued)*

Could the Change....	Yes/ No/ NA	Possible Effect	Action to Maintain or Improve Safety	Action For	By (Date)
Affect the involvement of direct supervision in the management of workplace injury and illness cases?					
Affect the manner in which safety performance data are collected, analyzed, and reported?					
Affect the manner in which safety performance goals or targets are established?					
Require revisions to written role descriptions for managers, supervisors, technical staff, operators, maintenance crafts, or other safety-critical personnel?					
Potentially affect the community?					
Necessitate discussions with the community about the change?					
Require changes in the way that drug and alcohol control policies are administered?					
Require changes to the safety training programs?					

TABLE B.2 Safety and Health Management Checklist *(Continued)*

Could the Change....	Yes/ No/ NA	Possible Effect	Action to Maintain to Improve Safety	Action For	By (Date)
Require changes to the safety training materials (i.e., workbooks, videos, etc.)?					
Require new or different instructors who must be trained?					
Result in changes in safety procedures, which will require new or additional training?					
Require training of existing employees in procedures or practices that will be new to them?					
Require changes in "on the job" training?					
Affect "hands on" computer application support?					
Result in changes to how the safety training recordkeeping system functions?					
Require special "one time" training to implement the change?					

TABLE B.3 Safe Work Practices Checklist

3 – SAFE WORK PRACTICES CHECKLIST					
Could the Change Require Changes in....	**Yes/ No/ NA**	**Possible Effect**	**Action to Maintain or Improve Safety**	**Action For**	**By (Date)**
Procedures or personnel involved in removing equipment from service or preparing it for maintenance?					
Blinding or isolation procedures?					
Cold work authorization procedures? (including procedures for notification of interconnecting units)					
Excavation procedures?					
Confined space entry procedure (reference EHS Regulatory Compliance Checklist)					
Lockout/tagout procedures? (reference EHS Regulatory Compliance Checklist)					
Hot work authorization procedures? (including Fire Watch and Vehicular Entry procedures)					
Hot tap procedure?					
Fire water use procedures?					
Flare opening procedures?					
Temporary utility connection procedures?					

TABLE B.3 Safe Work Practices Checklist *(Continued)*

Could the Change Require Changes in....	Yes/ No/ NA	Possible Effect	Action to Maintain or Improve Safety	Action For	By (Date)
Periodic monitoring of equipment? (e.g., operator checks or condition monitoring inspections)					
Safe Work Practices Auditing program?					
Pre-Job Planning and Hazard Control procedures?					

TABLE B.4 Process Safety Management/Risk Management Program (PSM/RMP) Checklist

4 – PROCESS SAFETY MANAGEMENT (PSM/RMP) CHECKLIST					
Could the Change Require Changes in PSM Programs for….	**Yes/ No/ NA**	**Possible Effect**	**Action to Maintain or Improve Safety**	**Action For**	**By (Date)**
Employee participation?					
Process safety information?					
Process hazard analyses?					
Operating procedures?					
Training?					
Contractors?					
Pre-startup safety review?					
Mechanical integrity?					
Hot work permit?					
Management of change procedures?					
Incident investigation procedures?					
Emergency planning and response?					
Compliance audits?					

TABLE B.5 Contractor Safety Checklist

5 – CONTRACTOR SAFETY CHECKLIST					
Could the Change Require Changes in....	**Yes/ No/ NA**	**Possible Effect**	**Action to Maintain or Improve Safety**	**Action For**	**By (Date)**
Persons responsible for implementing a contractor EHS program?					
New contractors in the plant?					
Types of work performed by contractors?					
Contractor prequalification procedures?					
Owner's EHS requirements for contractors?					
Owner's prebid package for contractors?					
Contractor selection process?					
Contractor training?					
Prejob activities?					
Contractor EHS statistical reporting?					
Audits of contractor work in progress?					
Investigation of contractor incidents?					
Contractor evacuation procedures?					
Periodic evaluation of contractor's safety and health performance?					

TABLE B.6 Emergency Response Checklist

6 – EMERGENCY REPONSE CHECKLIST					
Could the Change Require Changes in....	**Yes/ No/ NA**	**Possible Effect**	**Action to Maintain or Improve Safety**	**Action For**	**By (Date)**
The plant's written emergency response plans? (including fires, spills, and releases)					
The personnel who respond to emergencies as part of an organized response team?					
The personnel who respond to emergencies within a department?					
The plant's emergency alarm or notification system?					
The procedures for notifying off-duty personnel to respond to an emergency?					
The incident command system?					
Emergency response training?					
Personnel needing emergency response training?					

TABLE B.7 EHS Regulatory Compliance Checklist for Selected Regulations

7 – EHS REGULATORY COMPLIANCE CHECKLIST FOR SELECTED REGULATIONS					
Could the Change Require Changes in Compliance Programs for....	**Yes/ No/ NA**	**Possible Effect**	**Action to Maintain or Improve Safety**	**Action For**	**By (Date)**
OSHA (General)					
Injury/Illness Recordkeeping (29 CFR Part 1904)					
Handling OSHA inspections?					
OSHA General Industry Work Practices					
Lockout/Tagout (29 CFR 1910.147)?					
Confined Space Entry (29 CFR 1910.146)?					
Electrical Safety – General Industry (29 CFR Subpart S)?					
Electrical Safety - Special Industries (29 CFR 1910.269)?					
OSHA Emergency Response					
Emergency and Fire Prevention Plans (29 CFR 1910.38)?					
Fire Brigades (29 CFR 1910.156)?					
Hazardous Waste Operations and Emergency Response (HAZWOPER) (29 CFR 1910.120)?					
Medical Services and First Aid (1910.151)?					
OSHA PSM *(refer to separate checklist)*					

TABLE B.7 EHS Regulatory Compliance Checklist for Selected Regulations *(Continued)*

Could the Change Require Changes in Compliance Programs for....	Yes/ No/ NA	Possible Effect	Action to Maintain or Improve Safety	Action For	By (Date)
OSHA Other Selected Standards					
Scaffolds (29 CFR 1910.28 or 1926.450-454)?					
Powered Industrial Trucks (29 CFR 1910.178)?					
Excavation (29 CFR 1926 Subpart P)?					
OSHA Health Standards *(selected)*					
Hazard Communication (29 CFR 1910.1200)?					
Personal Protective Equipment (29 CFR 1910 Subpart I)?					
Respiratory Protection (29 CFR 1910.134)?					
Occupational Noise Exposure (29 CFR 1910.95)?					
Radiation (29 CFR 1910.96-97)?					
Asbestos (29 CFR 1910.1001 or 1926.1101)					
Benzene (29 CFR 1910.1028)?					
Lead (29 CFR 1910.1025 or 1926.62)?					
Access to Employee Exposure and Medical Records (29 CFR 1910.1020)?					
Bloodborne Pathogens (29 CFR 1910.1030)?					
Other Regulatory					

TABLE B.8 Occupational Health Checklist

8 – OCCUPATIONAL HEALTH CHECKLIST					
Could the Change Require Changes in....	**Yes/ No/ NA**	**Possible Effect**	**Action to Maintain or Improve Safety**	**Action For**	**By (Date)**
General					
Management of potential workplace illnesses?					
Administration of general medical examinations?					
Administration of OSHA-required medical examinations?					
Administration of OSHA health regulations?					
Exposure monitoring strategy?					
Handling employee complaints?					
Respiratory Protection *(reference EHS Regulatory Compliance Checklist)*					
Program administration?					
Persons subject to use of respirator?					
Fit testing program?					
Training?					
Selection and use?					
Cleaning, maintenance, and repair?					
Hearing Conservation and Noise *(reference EHS Regulatory Checklist)*					
Program administration?					
Persons subject to use of hearing protection?					
Audiometric testing & recording shifts?					

TABLE B.8 Occupational Health Checklist *(Continued)*

Could the Change Require Changes in....	Yes/ No/ NA	Possible Effect	Action to Maintain or Improve Safety	Action For	By (Date)
Training?					
Selection and use?					
Signage?					
Personal Protective Equipment *(reference EHS Regulatory Checklist)*					
Program administration?					
Persons subject to use of PPE?					
Selection criteria?					
Training?					
Hazard Communication *(reference EHS Regulatory Checklist)*					
Chemical-Specific Standards *(reference EHS Regulatory Checklist)*					
Heat Stress					
Administration of program?					
Engineering Controls					
Maintenance of ventilation systems?					

TABLE B.9 Process Unit Operability and Safety Effectiveness Checklist

9 – PROCESS UNIT OPERABILITY AND SAFETY EFFECTIVENESS CHECKLIST					
Could the Change Require Changes in....	**Yes/ No/ NA**	**Possible Effect**	**Action to Maintain or Improve Safety**	**Action For**	**By (Date)**
Human Factors					
Will the responsible operator be able to monitor critical controls and alarms?					
Will the responsible operator be able to deal with the number of alarms associated with an upset or emergency?					
Will the responsible console/board operator be able to monitor appropriately the number of control loops assigned?					
Are the roles, responsibilities, and authorities of the console supervisors and/or field operators affected by the change?					
Will the field operator be able to effectively monitor the operability and status/condition of the unit?					

TABLE B.9 Process Unit Operability and Safety Effectiveness Checklist *(Continued)*

Could the Change Require Changes in....	Yes/ No/ NA	Possible Effect	Action to Maintain or Improve Safety	Action For	By (Date)
Operating Procedures					
Have the operating procedures been revised to reflect the new assignments and duties, as appropriate?					
Do the procedures provide for safely conducting activities for each phase of operation? • Initial startup • Normal operations • Temporary operations • Emergency • Emergency contingency guidelines • Normal shutdowns • Startup following turnaround or emergency shutdown					
Are the procedures written such that staffing is appropriate for special tasks such as lighting furnaces, special line-ups and transfers, operator entry into confined spaces such as pits, or similar tasks where short-term assistance may be necessary?					
Emergency Procedures					
Can the operator(s) reasonably complete all tasks necessary to safely shut down the process?					

TABLE B.9 Process Unit Operability and Safety Effectiveness Checklist *(Continued)*

Could the Change Require Changes in....	**Yes/ No/ NA**	**Possible Effect**	**Action to Maintain or Improve Safety**	**Action For**	**By (Date)**
Does the answer change if all instrumentation fails simultaneously?					
Is staffing appropriate to make proper emergency communications (local and facility communications; affected units)?					
Does the operator have time to activate emergency systems such as manual sprinkler systems and fire water monitors?					
PHA Reviews					
Have the PHA reports been reviewed to identify events where safeguards include operator intervention and procedural activities involving operators?					
Are these safeguards still appropriate?					

Example B.3 Template MOPC Review and Approval Form

MOPC Title (Title of the position that is changing and the effective date.) NOTE: Action items below in most circumstances are completed after the change's effective date.
MOPC Coordinator (Person who is responsible for the MOPC review and MOPC documentation including documentation of completed action items. This is typically the supervisor or exiting employee of the position that is affected.)
Description of Change (Document here why you are making this change.) Examples: Replace retiring employee X with existing employee Y from department Z. New responsibilities added to position X requiring new skills/competencies.
Instructions – Review the items in the checklist in Table B.10 to identify actions required to manage this personnel change appropriately. Not all items might apply. Document actions, who will complete them, and the due date. –Actions typically include training, document reviews, updating lists, updating documents, communications, etc. Some actions are required by the employee (e.g., review document X). Some actions are required by other employees (e.g., safety department administrator to update the callout list). –It might be necessary to identify an interim action until the final action is complete (e.g., The environmental manager will assume crisis notification responsibilities until the new safety manager is onboard).
The MOPC review will occur as soon as possible after the position change effective date. **Supervisor signature (copy of email response is acceptable):**___________________ **Date:**____ **The MOPC coordinator or supervisor is responsible to ensure this completed form (including indication of completed actions) is retained per PSM MOC corporate records retention requirements.**

TABLE B.10 Management of Personnel Change (MOPC) Checklist

Responsibility or Action	Action Required and by Whom:	Due Date:
Involved in key decision making and/or communications, particularly during off-hours (i.e., call-outs, flare opening, hot tapping, unit shutdowns, corporate reporting for crisis management, site's emergency notification system, etc.)		
Knows the location and proper response to emergency situations (i.e., emergency shutdown systems, evacuation points, etc.)		
Maintains, reviews, or certifies operator/craft procedures or training		
Participates within Emergency Response Teams or Incident Command System		
Responsible for process safety information; including safe operating limits, and understands where information is stored and how it is maintained and accessible to employees		
Participates in MOC coordination, administration, EHS/Technical reviews, PSSRs, or approvals (approval to commission)		
Needs general understanding of MOC applicability		
Participates in unit PHAs (HAZOPs), is responsible for PHA action items or communicating PHA information to employees		
Has identified accountabilities or responsibilities under the Mechanical Integrity System		
Mechanical integrity checks (critical alarms, equipment, monitoring safety system checks, etc.)		
Participates in safety committees or other EHS improvements		
Is a technical or experienced resource for the operating personnel on the process		
Needs understanding of safety procedures and/or industrial hygiene standards (identify specific ones)		
Responsible for regulatory compliance		

TABLE B.10 Management of Personnel Change (MOPC) Checklist *(Continued)*

Responsibility or Action	**Action Required and by Whom:**	**Due Date:**
Has identified accountabilities or responsibilities under environmental procedures or processes		
Responsible for notifying corporate or regulatory agencies of safety/environmental incidents		
Conduct safety and/or environmental audits or incident investigations		
Conduct job task observations		
Coordinate projects		
Coordinate contractor work affecting PSM, safety, health, environmental, or mechanical integrity		
Other PSM, safety, or environmental tasks		
Employee's position charter adequately reflects all key safe operation, emergency response, process safety, mechanical integrity, and safety, health, and environmental job responsibilities		
Other actions needed to manage this personnel change		

Example B.4 Organizational Change Risk Assessment Procedure

1.0 PURPOSE

1.1 The purpose of this procedure is to define the process that ensures that proposed organizational changes receive appropriate risk assessment evaluation.

1.2 The intention is to evaluate the potential impact of the organization's ability to control accident hazards and to ensure that there are no adverse effects on the organization's ability to deliver quality products and services to its customers.

2.0 SCOPE

2.1 This procedure applies to proposed and imposed organizational change (roles, responsibilities, terms and conditions of employment), including position elimination/addition, personnel replacement, changes to hours and conditions.

3.0 REFERENCE AND LINKS:

4.0 DEFINITIONS

4.1 **Relevant Change** – A relevant change within the organization is one that could potentially have an impact on the organization's ability to control accident hazards, or one that could have a significant impact on the organization's ability to manufacture and deliver high-quality materials to its customers. As a guideline, the following may be considered a relevant change:

4.1.1 An established position is removed from the organization.

4.1.2 A new position is added to the organization.

4.1.3 Individuals are required to take on new responsibilities demanding skills and competencies unconnected with those previously required.

4.1.4 Individuals relinquish responsibilities for tasks without those tasks being reallocated.

4.1.5 Changes to terms and conditions of employment such as hours, shifts, locations, etc.

4.2 Approver – This is the site management team member in whose team the change is proposed.

5.0 PROCEDURE

5.1 The approver shall, upon determining that a proposed change is a relevant change, complete the Organizational Change Initiation Form.

5.2 The approver shall assemble a team to complete the risk assessment.

5.2.1 The team shall consist of at least two employees. As many team members as necessary should be selected to ensure identification of duties relevant to the safe operation of the plant and the production of quality product.

5.2.2 In selecting the team, a mix of attributes must be achieved as appropriate to the change under consideration. The team shall consider:

5.2.2.1 Manager of the department undergoing change

5.2.2.2 Customers of the position/organization undergoing change

5.2.2.3 The employee(s) involved in the change

5.2.2.4 Co-workers of the employee/positions undergoing change

5.2.2.5 EHS/Regulatory affairs manager

5.2.2.6 Process safety manager

5.2.2.7 Adequate seniority/experience

5.2.2.8 Adequate understanding of human factors

5.2.2.9 Adequate technical understanding and experience with the issues involved

5.2.2.10 Adequate understanding of organizational and business issues involved in the change

5.2.2.11 Adequate understanding of the concept of risk

5.3 The team shall use its collective knowledge and experience to identify relevant duties performed by personnel/positions undergoing change. This phase of the risk assessment shall be documented.

5.3.1.1 Duties identified in the position job brief(s)

5.3.1.2 Duties identified by the relevant manager

5.3.1.3 Duties identified by the position's customers

5.3.1.4 Employees who held the positions involved in the change

5.3.1.5 EHS, responsible care, regulatory and process safety-related duties

5.3.1.6 Note: Ancillary duties, while not identified in a job brief, may be vital to the functioning of the organization. For example, an existing project engineer, who previously held a process engineering position, may have continued to provide valuable process engineering expertise after transfer to the new position. This may have occurred despite having been transferred to the project engineering position, without being identified in the project engineer job brief. Should this employee later leave the company, the replacement hired to fill this position may not possess the skills necessary to provide process engineering expertise, resulting in a loss of technical expertise to the organization.

5.3.2 Wherever required, the team should consult as widely as necessary to allow them to come to an informed decision regarding the risks associated with the change.

5.3.3 The team may find the guide words below helpful to consider during the risk assessment.

- Competence
- Communication
- Experience
- Knowledge
- Workload

- Stress
- Quality of Product
- Integrity of Product
- Integrity of Records
- Budget
- Staff Appraisal
- Authorization Level
- Authority
- Work Force Reaction
- Morale
- Logistics
- Hours Overtime
- Location

5.4 After identification of relevant tasks performed by the existing organization, the team shall identify how each of these functions will be performed following implementation of the change. This shall be documented.

5.5 Next, the team shall determine, by consensus, what risks may be presented by the change.

Risks may take any form, including but not limited to:

5.5.1 Duties/responsibilities that exist in the current organization that are not clearly assigned in the new organization.

5.5.2 Duties/Responsibilities for which the position holders under the new organization are not adequately trained or do not have sufficient experience to perform adequately.

5.5.3 Loss of organizational knowledge and technical expertise.

5.6 Risks identified by the team shall be documented. For each risk identified, the team shall document the issue, risk, and any risk mitigations proposed by the team.

5.7 The approver shall then identify the risk mitigations to be implemented, the responsible person for implementing each risk mitigation, and the target date for completion.

- **5.7.1** The approver shall justify in writing—attached to the risk assessment, any risks for which mitigation will not be implemented.

5.8 The completed change documentation, risk assessment, and justification for risks not mitigated shall be submitted to the relevant vice president for approval.

5.9 Due to the sensitive nature of certain organizational changes, the finalized assessment may not be completed until after the change is implemented. These situations must be managed by the relevant department manager and the assessment completed within a reasonable period (<30 days) from implementation.

5.10 The approver shall document completion of each of the risk mitigation action.

- **5.10.1** The risk assessment forms shall be forwarded to the process safety manager after all risk mitigations have been implemented.

5.11 The process safety manager shall maintain the completed risk assessment forms for a period of five years.

Example B.5 Organizational MOC Procedure to Assess Minimum Process Unit Staffing Levels

Organizational Management of Change

Assess Minimum Process Unit Staffing Levels To Meet Process Safety Requirements

1. **Purpose**

   The purpose of this document is to provide guidance to assess the minimum process unit staffing levels necessary to meet process safety requirements under various emergency conditions. This guidance is designed to evaluate minimum process unit staffing level requirements during emergency operations (a.k.a. safe-off) but does not address routine and planned operations. It is intended to provide a consistent methodology for conducting staffing assessments. Any proposed staffing changes shall be managed via the MOC procedure.

2. **Assumptions**

   The following assumptions are used throughout this document. These assumptions shall be validated by the review team.

   a. This document focuses on the required duties of operations staff during emergency operations.
   b. Critical instrument systems [such as critical alarms, trips, mechanically operated valves (MOVs), interlocks, PLCs, mitigation equipment, critical corrective actions, etc.] integral to the emergency procedures, and other unit upsets or emergency situations, are tested at an appropriate frequency to ensure high reliability.
   c. Emergency procedures are current and up-to-date, and the current operating staff has demonstrated the capability to execute the procedures effectively.

d. Operator staffing requirements for restarting a unit from a partial and/or complete emergency shutdown will be augmented as necessary.

3. **<u>Conducting the Review</u>**
   a. <u>Selection of Review Team Members</u>
      i. One or more persons who have experience and knowledge specific to the process unit operations being evaluated.
      ii. A technical person who has expertise in engineering, process unit operations, and/or unit process control.
      iii. A team leader, who is knowledgeable in process safety. Team leader consistency for multiple reviews within the site is deemed important.
   b. <u>Preparation Materials</u>
      i. Emergency procedures for the following scenarios:
         - loss of utility (steam, instrument air, process air, nitrogen, cooling water, fuel gas, power, etc.)
         - loss of control
         - loss of feed
         - loss of product outlet
         - loss of hydrocarbon containment
         - other emergencies

      ii. Job descriptions/task analysis/job safety analysis (if available) for all operator jobs
      iii. Process unit emergency response plans (if available)
      iv. List (if available) of critical alarms, trips, interlocks, critical corrective actions and other critical instrument systems to validate reliability

v. Validation of "Assumptions" (Section 2)
vi. Ten-year serious incident history for the process unit (plus similar process units at other locations, if available)

c. Review Team Methodology
i. Complete a tabletop review of all prerequisite materials.
ii. Complete the emergency procedures review, and determine the minimum unit staffing level for the emergency case (Section 4).
iii. Complete field observations and unit walkthrough (if appropriate).
iv. Document any technology improvements or equipment additions/changes identified by team members that could facilitate the emergency cases.
v. Prepare final document.

4. Emergency Procedures

a. Definition

For the purposes of this assessment, the emergency is determined to be complete when energy input has been removed from the process unit, major equipment has been secured from damage, and the environment is protected from an uncontrolled hazardous release. The process unit may not be totally secure at this point; however, it is considered safe. In some cases, this definition may reduce the primary emergency shutdown requirements.

b. Asset Management Staffing

For each defined emergency scenario, the team must:
i. Review the emergency tasks and remove any steps not required to meet the definition of a safe state (Section 4.a).
ii. Prioritize the remaining tasks, and identify sequential and parallel paths.
iii. Review the geographic layout of the unit relative to sequential and parallel paths.

iv. Time reference the tasks for both sequential and parallel paths. Consideration must be given to the physical demands of the procedural steps (e.g., closure of hard to operate manual valves) and weather factors (e.g., address worst-case winter conditions).
v. Determine the minimum staffing required to secure a safe state within 30 minutes with existing shutdown systems or trips.

For the loss of containment scenario, the team shall identify all major process unit incidents over the last 10 years and determine the minimum staffing required to shutdown the process unit in the limiting case. When evaluating these loss of containment scenarios, it is noted that the facility response should be per the HAZWOPER standard and the process unit emergency response plan. It should be recognized that trained emergency responders will aid in securing the facility as directed by process unit personnel.

Attachment A shall be used to summarize the emergency procedure reviews.

5. **Summary Report**

   A written report (with attached preparation materials) shall be prepared by the review team and contain the following:

   a. A discussion of the minimum staffing required to execute the limiting emergency case
   b. Discuss any staffing options
   c. Document recommended action items from review

ATTACHMENT A

Emergency Procedures Review Summary

Emergency Procedure	Manpower Required	Comments
1. Loss of steam		
2. Loss of instrument/process air or nitrogen		
3. Loss of cooling water		
4. Loss of fuel gas		
5. Loss of electric power		
6. Loss of control		
7. Loss of feed		
8. Loss of product outlet		
9. Loss of hydrocarbon containment		
10. Other process unit-specific emergency procedures		
LIMITING CASE		

REFERENCES

Canadian Society for Chemical Engineering, *Managing the Health and Safety Impacts of Organizational Change*, Ontario, 2004.

Chemical Manufacturer's Association, *Management of Safety and Health During Organizational Change*, Washington, DC, 1998. This organization is currently known as the American Chemistry Council.

INDEX

Acquisitions, Mergers and Joint Ventures, 127

Activity mapping, 46

Baker Panel, 124

Baker Panel report, 137, 141

Bayer CropScience, 87, 103

Bhopal, India, 11, 83, 129

Bow Tie, 45

BP, 21, 107, 124, 137

CCPS, 11, 12, 14, 15, 24, 25, 26, 54, 60, 62, 128, 130, 146. *See* Center for Chemical Process Safety

Center for Chemical Process Safety. *See* CCPS

Centralization, 114, 115, 116

Checklist, 43, 109, 120

Chemical Safety and Hazard Investigation Board. *See* CSB

Closeout, 28, 32, 61

Competencies, 38

CSB, 73, 104. *See* Chemical Safety and Hazard Investigation Board

Decentralization, 114

Downsizing, 83, 85, 99, 120

DuPont, 135, 137

Emergency, 74, 85

Emergency change, 57, 94

Emergency procedures, 199

Emergency response, 35, 43, 45, 73, 95

Esso, 65, 115

Flixborough, 11

Grangemouth, Scottland, 21

Health and Safety Executive, 13, 44, 189

Hickson & Welch, 1

Hierarchy changes, 113

Implementation plan, 56, 60

Indicators, 59

Institute, West Virginia, 87, 103

Job competency, 102

Ladder assessment, 194

Layoff, 83

Location, 68

Longford, Victoria, Australia, 65, 115

Mergers and acquisitions, 15

Mission and vision statements, 136

MOC, 12, 13, 15, 17, 24, 27, 28, 29

Organizational change, 29

Personnel change, 30, 83, 93, 199, 200

Personnel turnover, 95

Policy changes, 135, 144

Reorganization, 113, 117

Risk assessment, 27, 32, 35, 46, 52, 54, 57, 76, 106, 110, 199

Scenario assessments, 44

Shift cchedules, 69

SME, 118. *See* Subject Matter Expert

Span of control, 121, 122

Strikes, 93

Subject matter expert. *See* SME

Task allocation, 99, 106

Task mapping, 45, 110, 120, 129

Temporary changes, 61, 93, 106

Texas City, Texas, 73

Transition plan, 58

Turnaround, 74, 75, 79, 144

Union Carbide, 83, 129

West Yorkshire, England, 2

Whiting, Indiana, 107

Working conditions, 65

Printed and bound by CPI Group (UK) Ltd, Croydon, CR0 4YY

23/04/2025

14660906-0003